URBAN TRANSPORT SYSTEMS

Edited by **Hamid Yaghoubi**

Urban Transport Systems
http://dx.doi.org/10.5772/62814
Edited by Hamid Yaghoubi

Contributors
Teresa Moreno, Ammar AbuLibdeh, Milos Poliak, Juan Moreno Garcia-Loygorri, Jose Riera, Carlos Rodriguez, Tamer Cetin, Jaime Santos-Reyes, Vladimir Avalos-Bravo, Diego Padilla-Perez, Qi Li, Weirong Chen, Maja Ahac, Stjepan Lakušić, C.-Y. Cynthia Lin Lawell, Justin Beaudoin, Mihaela Grubišić Šeba, Konstantinos Vogiatzis, Georges Kouroussis

CBS, Edition **2018**

Published by InTech
Janeza Trdine 9, 51000 Rijeka, Croatia

Publishing Process Manager Martina Usljebrka

Technical Editor SPi Global

Cover Designer InTech Design Team

Additional hard copies can be obtained from orders@intechopen.com

Urban Transport Systems , Edited by Hamid Yaghoubi
p. cm.
Print ISBN 978-953-51-2873-1
Online ISBN 978-953-51-2874-8

Contents

Preface

The book *Urban Transport Systems* contains a collection of latest research developments on the urban transportation systems. This book describes rail transit systems, subways, bus rapid transit (BRT) systems, taxicabs, automobiles, etc. It also studies the technical parameters and provides a comprehensive overview of the significant characteristics for urban transportation systems, including energy management systems, wireless communication systems, operations and maintenance systems, transport serviceability, environmental problems and solutions, simulation, modelling, analysis, design, safety and risk, standards, traffic congestion, ride quality, air quality, noise and vibration, financial and economic aspects, pricing strategies, etc. This professional book as a credible source can be very applicable and useful for all professors, researchers, practitioners, students and others interested in urban transportation systems.

It consists of 11 chapters. Chapter 1 models track gauge degradation on small urban rail networks. Track gauge is the most significant tram track geometry parameter. Its degradation, which manifests as gradual increase of gauge deviation from prescribed values during track exploitation, causes poor ride quality, reduces safety and triggers most of the maintenance activities. To optimize tram track maintenance procedures, it is necessary to increase the proportion of preventive maintenance at the expense of corrective maintenance. This requires creation of physical model of track degradation. Conducted survey of conventional track degradation models showed that, in order to quantify the influence of track design, construction and exploitation characteristics on gauge degradation, it is most favourable to adopt the mechanistic-empirical modelling approach. Zagreb high-capacity tram network presents an optimal testing ground for exploration of the possibilities for tram track gauge degradation model development. Analysis of modelling results gave new, practical insights into the effects of tram track design and construction elements and exploitation characteristics on gauge degradation. These models represent the first step towards predictive maintenance system establishment on Zagreb tram tracks.

Chapter 2 discusses a hybrid power system which consists of multiple proton exchange membrane fuel cell (PEMFC) systems, batteries and supercapacitors (SCs) that is developed for a hybrid tramway. Three energy management strategies which included a fuzzy logic control (FLC), an equivalent consumption minimization strategy (ECMS) and a state machine strategy based on droop control (SMS-DC) are utilized to coordinate multiple power sources, avoid the transients and rapid changes of power demand and achieve high efficiency without degrading the mechanism performance for an energy management system of hybrid tramway. A hybrid system model of tramway is established with commercially available devices, and then the different energy management strategies are evaluated with a real driving cycle of tramway from Turkey. The results compared with these strategies demonstrate that the higher average efficiencies of the tramway, the lower tramway equivalent hydrogen consumptions and more efficient use of the batteries and SC energy are achieved by the SMS-DC. Therefore, the appropriate energy management system for high-power hybrid tramway will improve the hydrogen consumptions of overall hybrid system and the efficiencies of each power source.

Chapter 3 describes the main features of the wireless communication systems of urban rail and related applications. The perspective will be complete: application, network and physical layers will be discussed. Moreover, to properly address some of the challenges that need to face these systems, the chapter will provide a deep insight into propagation issues related to tunnels and urban areas. Finally, a detailed survey on the directions of research on all these topics will be provided.

Chapter 4 evaluates airborne and ground-borne noise and vibration for urban rail transit systems. The environmental effect of ground-borne vibrations and noise generated by urban rail transit systems is a growing concern in urban areas. This chapter will review, synthesize and benchmark new understandings related to railway vibrations and associated airborne and ground-borne noise. The aim is to provide new thinking on how to predict noise and vibration levels from numerical modelling and from readily available conventional site investigation data. Recent results from some European metropolises (Brussels, Athens, etc.) are used to illustrate the dynamic effect of urban railway vehicles. It is also proved that train type and the contact conditions at the wheel/rail interface can be influential in the generation of vibration. The use of noise mapping–based results offers an efficient and rapid way to evaluate mitigation measures in a large scale regarding the noise exposure generated by dense urban railway traffic. This information provides assistance to future researchers attempting to simulate railway vehicle vibrations and noise.

Chapter 5 describes air quality sampling campaigns in three European subway systems (Barcelona, Athens and Oporto) that were conducted in order to characterize particulate matter (PM) to better understand the main factors controlling it. PM mass concentrations varied among the European subway platforms and also within the same underground system, this being mainly associated to differences in the design of the stations and tunnels, system age, train frequency, ventilation and air conditioning systems, commuter's density, rails geometry and outdoor air quality. PM concentrations displayed clear diurnal patterns, depending largely on the operation and frequency of the trains and the ventilation system. Chemically, subway PM 2.5 on the platforms consisted of iron, carbonaceous material, crustal matter, secondary inorganic compounds, insoluble sulphate, halite and trace elements. Fe was the most abundant element, accounting for 19–46% of the bulk PM 2.5 which is generated mainly from mechanical wear at rail-wheel-brake interfaces. A source apportionment analysis allowed the identification of outdoor (sea salt, fuel oil combustion and secondary aerosol) and subway sources on platforms. The use of air conditioning inside the trains was an effective approach to reduce exposure concentrations, being more efficient removing coarser particles. PM concentrations inside the trains were greatly affected by the surrounding (i.e. platforms and tunnels) air quality conditions.

Chapter 6 reviews the latest theoretical and empirical research on the effects of public transit investment on congestion and the demand for automobile travel and air quality. Traffic congestion is ubiquitous across urban roadways, and the adverse health effects accompanying deteriorating air quality are an ongoing concern. Beyond these local effects, transportation is also a major contributor of greenhouse gas emissions and is thus a significant element of the climate change debate. A contentious issue currently confronting transportation analysts and policymakers is what the effects of public transit investment on traffic congestion and on air quality are and therefore what the appropriate level of public transit investment should be. While public transit receives plenty of political support for its "green" reputation and its contribution to sustainability, there have been relatively few studies examining the ex post effects of public transit investment on traffic congestion or air quality.

Chapter 7 presents a statistical analysis of historical data of BRT lines A and B accidents that have occurred in Mexico City from 2005 to 2015. Some of the key conclusions are the following: (a) 484 accidents have occurred when considering both lines A and B. The most critical years have been 2008, 2011 and 2012; the least critical year, on the other hand, has been 2010; (b) overall, the frequency of accident occurrence has been decreasing in both lines; (c) the most critical seasons of the year have been the following: autumn (27.7% in line A) and winter (32% in line B); (d) the frequency of accidents increases when approaching the end of the week (Thursday and Friday) and the frequency of accidents decreases sharply at weekends; (e) 48.28% and 54.47% of accidents have occurred at the three peaks (i.e. morning, afternoon, evening/night) in lines A and B, respectively; (f) 64.8% (22/73) of pedestrians have been killed when collided with the BRT buses; and (g) the most critical section of the BRT lane has been identified with 38 (11.87%) accidents and for the case of line A. Future work includes the upgrade of the analysis by considering missing data related to the fatalities. Further, an accident analysis is essential to understand the multicausal factors leading to the accidents; finally, some relevant statistical significance tests on the data are needed.

Chapter 8 describes a procurement process for providing transport serviceability by public passenger transport. The objective of the chapter is to present individual steps for procurement of public transport services. These steps consist of identification of objectives, definition of requirements for transport serviceability, risk allocation between contractual parties, drafting a public service contract and a process of selecting a service operator. Special attention is paid to the risks and their influence on contracting parties. The chapter also characterizes procedure for the direct award of a public service contract, that is without competitive tendering. The chapter tries to define the impact of the direct award of contracts on the scope of services provided.

Chapter 9 considers financial aspects of urban transportation systems. The demand for urban transport quality is typically above financing possibilities of public authorities. Financing urban transport has always been one of the prime problems of city authorities because of the necessity to connect the city centres with their surroundings and to enable time-saving and thus better quality of life for the citizens. This problem especially emerges in the twenty-first century when the citizen requirements for better urban and suburban connectivity are coupled with smart, intermodal and energy-efficient urban transport. The financing problem of urban transport is somewhat simpler in very populated and developed areas as the growing number of public transport users continuously finances urban transport fleet renewal. However, less developed areas have to have integrated pricing and social policies towards the end users of urban transport which often turn to be unsustainable in the longer period of time. Depending on the project size, financial strength of municipalities and/or central state, urban transport infrastructure construction and maintenance are typically financed from national or local state funds or borrowing. Some urban transport lines can also be given in a concession. Financing urban transport encompasses either financing urban transport infrastructure construction or financing fleet renewal or combined financing of both urban transport infrastructure and fleet renewal. The EU funds have contributed much to financing urban transport needs, especially in large metropolitan areas. Yet, many countries opt for financing regional and cross regional connectivity by roads, rail, airports or waterways, while urban transport remains a care of national or local public authorities. Most literature is devoted to rail, road and port infrastructure construction in general, while urban transport fleet renewal and operating performance of urban transport operators have not been widely discussed. This chapter aims to partly fill in this gap for the selected cities of formerly planned economies of the Central and Eastern and Southern and Eastern Europe.

Chapter 10 examines the rise of ride-sharing in urban transportation and its effect on the traditional taxicab services. The aim is to evaluate whether ride-sharing is a threat or an opportunity for the future of urban transportation. The rise of ride-sharing has influenced the nature of urban transport in many different areas from the quality of service to the structure of prices and demand within public transportation. The uber-experience suggests that ride-sharing in taxi markets has become substitution for the traditional cab services. Clearly, the ride-sharing system competes with the traditional taxicab service. The success of ride-sharing stems from its impact on consumer welfare. This refers to economic efficiency considerations under competitive market conditions. Most importantly, this system eliminates the transaction costs of a taxicab market under regulated market conditions. Conversely, the failure of traditional taxicab services stems from distortions inherent in a regulatory process. Price controls and entry regulations lead to artificial monopoly rents through medallion prices and taxi fares. Clearly, while ride-sharing increases consumer welfare, the strictly regulated taxicab markets lead to deathweight loses in consumer welfare and economic inefficiency. For that reason, in order to better understand the effect of ride-sharing as a transport mode in urban transportation on the traditional taxi services, we also have to take into consideration the reasons and results of taxicab regulation. For this aim, the chapter provides an overview of the literature on the regulation of taxis. This section of the chapter includes the economic rationale for regulation in the traditional taxicab markets, the lessons from some regulatory experiences and the effect of fare controls and entry restrictions in practice. The chapter also discusses the impact of ride-sharing on the cab service and its regulation in the traditional taxi markets. It concludes with a critical statement on regulation, deregulation and competition in taxi markets, including some policy suggestions. In this context, the chapter consists of three main sections including the introduction. In section two, the rise of ride-sharing and its effect on taxi services in urban transportation are evaluated. In section three, the regulation of taxicabs is discussed and some policy suggestions about regulation, competition and deregulation are introduced by taking into account the rise of ride-sharing.

Chapter 11 reviews traffic congestion pricing. Road traffic congestion is recognized as a growing and important urban ill. It occurs in different contexts, takes on many faces and is caused by a variety of processes. It affects both work trips and non-work trips and both passengers and goods flow. It affects the quality of life and the competitiveness of a region. It is an attendant cost that arises in the forms of delay, environmental degradation, diminished productivity, standard of living and wasted energy. Congestion pricing can result in winners and losers among different socio-economic groups. However, different studies differ in their conclusions about who wins and who loses because of different assumptions made. This chapter reviews the concepts of congestion pricing as a mitigation policy to reduce road congestion and reviews the concept of equity. This chapter aims to provide theoretical research that enhances our understanding of congestion pricing policy and the equity implications of this policy.

Dr. Hamid Yaghoubi

Director of Iran Maglev Technology (IMT), Iran

Chapter 1

Track Gauge Degradation Modelling on Small Urban Rail Networks: Zagreb Tram System Case Study

Maja Ahac and Stjepan Lakušić

Additional information is available at the end of the chapter

http://dx.doi.org/10.5772/65086

Abstract

Track gauge is the most significant tram track geometry parameter. Its degradation, which manifests as gradual increase of gauge deviation from prescribed values during track exploitation, causes poor ride quality, reduces safety and triggers most of the maintenance activities. To optimize tram track maintenance procedures, it is necessary to increase the proportion of preventive maintenance at the expense of corrective maintenance. This requires creation of physical model of track degradation. Conducted survey of conventional track degradation models showed that, in order to quantify the influence of track design, construction and exploitation characteristics on gauge degradation, it is most favourable to adopt the mechanistic-empirical modelling approach. Zagreb high-capacity tram network presents an optimal testing ground for exploration of the possibilities for tram track gauge degradation model development. Analysis of modelling results gave new, practical insights about the effects of tram track design and construction elements and exploitation characteristics on gauge degradation. These models represent the first step towards predictive maintenance system establishment on Zagreb tram tracks.

Keywords: tram track, gauge degradation, influential factors, mechanistic-empirical model, segmented database, regression analysis

1. Introduction

In the past few decades, a growing number of cities are turning again to the tram as an efficient, adaptable and environmentally friendly mean of urban public transport. Numerous planned projects for renovation of existing, revitalization of historical and construction of new tram lines prompted the International Association of Public Transport (UITP) to, referring to the experience of the rail intercity systems, warn about the following fact. The introduction of modern tram systems design solutions based on new technologies and construction materials in the long run will not yield the desired effect of establishing a sustainable urban transport system. This can only be achieved by the simultaneous development of modern tram infrastructure management system based on networking and mutual complementarity of design and maintenance activities in integrated tram track maintenance planning model [1].

Rail system management is a complex interdisciplinary process that includes managing, operation and maintenance of rail infrastructure and rolling stock, and organization of rail transportation. The emphasis that is nowadays placed on the necessity of reducing the total costs of rail systems requires optimizing each management step, including the track maintenance. Since the high implementation costs initiate significant investments in systems maintenance, primary objective for rail infrastructure management is to ensure a safe and comfortable ride with as long as possible system exploitation without any maintenance.

As rail vehicles of different weights run on tracks with various speeds, a wide range of stresses occur in the elements of track superstructure. They are a consequence of vehicle loading and the dynamic forces such as centrifugal, braking, acceleration, hunting oscillation, vertical inertia and vibration forces caused by irregularities in the rail and wheel contact surface. The consequences of these, usually very large, forces on track elements are numerous. Formation of rolling contact fatigue cracks, plastic flow deformations, shelling and uneven wear of rails, failure of rail fastening elements and changes in track geometry are the main adverse phenomena during track exploitation. They need to be adequately recognized and addressed by carrying out appropriate maintenance work.

The overall process of railway track maintenance includes the two basic types of activities—inspection and intervention.

Inspections are preformed in order to acquire information needed for defining the maintenance schedule. These information can be gathered manually (with the use of analog and digital measuring instruments, usually when small sections of track are to be inspected) and automatically (with the use of inspection car) [2].

Interventions comprise of corrective and preventive track maintenance. *Corrective maintenance* is carried out in the case of detecting defects on the track [3]. *Deferred corrective maintenance* is applied to small defects that do not require immediate action, but they are rather grouped and treated subsequently. *Immediate corrective maintenance* is applied in case of major track defects that require immediate intervention to ensure traffic safety [4, 5]. *Preventive maintenance* activities are preformed to maintain correct and reliable operation of the system. Their goal is to prevent and eliminate the cause of the defect. They can be *planned* under constant sustainability (predefined date) or *predictive* according to the system condition. They include monitoring of the tracks parameters and elements condition and behaviour, and development of methods, techniques, and equipment for the systematic system monitoring. [3].

To optimize track maintenance procedures, it is necessary to increase the proportion of preventive maintenance at the expense of corrective maintenance. This can be achieved by creation of a rail track *maintenance planning models*.

http://dx.doi.org/10.5772/65086

Section 2 gives an overview of general structure, classification and examples of several models developed for the conventional track systems. Section 3 gives an overview of the physical tram track gauge degradation model creation—from the influential factors and modelling approach identification to the assessment of the model representativeness. Section 4 presents results of model analysis in the form of the observed tram track gauge degradation influential factors ranking by the level of their impact on the degradation rate. Section 5 gives concluding remarks and recommendations for future work.

2. Track system maintenance planning models

Maintenance planning model is a decision-making support tool for systems maintenance management. It is used to assess the impact of the maintenance work implementation on the system service life. A comprehensive maintenance planning model consists of one or more *physical models*, integrated into *economic model* for planned maintenance activities costs estimation. Physical models can be used to describe exploitation behaviour of systems individual components or system as a whole. Activities justified by maintenance planning model need to be consolidated into the strategic maintenance plan taking into account time constraints and available resources for their implementation.

Creating a physical model to determine the exploitation life of the system's individual element, or the system as a whole, is the first step of establishing a modern approach to maintenance planning. This model can be established as a *failure* or *degradation model*, depending on which element and the type of maintenance work it describes.

Failure is defined as the point in time after which it is necessary to replace the observed element. It can be interpreted as actual physical failure or as economic failure that is a point in time after which is economically viable to replace the element. To establish a failure model, it is necessary to have information on the exploitation conditions in which the final element failure will occur. This information can be collected by long-term field observations or laboratory tests, by subjecting the elements to accelerated loads that may arise during the system operation.

Degradation is defined as gradual reduction of the observed elements quality, until it loses the properties required for whole systems quality. Since service life of rail systems is measured in decades, developed physical models of their exploitation behaviour rely on degradation monitoring approach. Mathematical models of track degradation estimate loss of tracks or its individual elements quality. If it is too low to ensure prescribed traffic safety and comfort, model suggests implementation of certain maintenance activities.

2.1. Rail track degradation models classification

Systematic research of rail track degradation began during the 1980s and 1990s of the past century, when the availability of data on tracks exploitation behaviour, especially in a digital format suitable for detailed analysis, was very modest. For this reason, researches mostly resulted in simple physical models of superstructure elements (rails, fastenings, sleepers and ballast) degradation [6].

A literature review showed that more recent studies, although numerous, are exclusively limited to standard 1435 mm gauge ballasted tracks, used for intercity traffic. To narrow gauge tram tracks, as a special group of slab track structures, is generally devoted less attention than to conventional ballasted railway tracks. Also, the existing railway regulations (International Union of Railways (UIC), Verband Deutscher Verkehrsunternehmen Oberbau-Richtlinien, European Committee for Standardization (CEN)) in general do not consider tram tracks [7]. Because of that, the following overview of degradation models and their classification is based exclusively on data for conventional ballasted intercity rail track systems.

Physical models of track degradation can be classified given the object of modelling, level of detail of the model and the methods of track data collection and processing during the model development.

Given the *object of modelling*, there are models of track geometry degradation and models of permanent way elements degradation (rails, sleepers, fastenings and ballast) [8].

Modelling the *track geometry* degradation refers almost exclusively to the modelling of track vertical geometry degradation, which occurs during exploitation due to settling of ballast material below the sleepers. There are two reasons for this. Geometry degradation rate is larger in the vertical than in the horizontal track plane [9]. Therefore, the vertical geometry is more relevant for the maintenance periodicity determination. The second reason is that modelling of the horizontal track geometry is substantially more complex task because it can degrade in either or both directions [10].

Most *permanent way elements* degradation models deal with the problem of rail wear. They usually state that rail wear at a given point of the track is proportional to the energy loss due to rail-wheel friction, and this energy loss is proportional to the hauled tonnage [11].

Given the *level of detail* and application of results, the analyses carried out during track degradation models creation can be divided into two basic groups.

Micro-analyses result in detailed *microscopic* track degradation models. These models are dealing with the forces on the individual permanent way elements and are usually based on engineering judgment or empirical evidence. Micro-analysis is performed through the analysis of the geometric parameters graphical representations and is used for occasional interventions with the purpose of eliminating defects on tracks.

Macro-analyses involve track data statistical analysis in order to determine the patterns in track behaviour during exploitation. They result in *macroscopic* track degradation models. These models, by inferential or descriptive statistical analysis of the measured track data, establish a correlation between the track degradation rate, exploitation conditions (exploitation intensity, speed and vehicle type) and used track superstructure elements. They are commonly used for cost-benefit analysis of rail infrastructure management strategies and contribute to defining the general policy of funds allocation and implementation of track maintenance [12].

http://dx.doi.org/10.5772/65086

2.2. Review of track degradation models

It is in the nature of the railway industry to hinder publicly publishing a detailed description of the track exploitation behaviour modelling procedures. Therefore, this review shows only a number of model examples (**Figure 1**) listed in the publicly available literature. Reviewed researches on conventional track structures degradation were carried out by the individual national railway administrations as well as universities and other research institutions. The leading institutions in this research area are Transportation Technology Center of Association of American Railroads (TTCI, Colorado), European Rail Research Institute (ERRI, the Netherlands), Queensland Rail (QR, Australia), Railway Technical Research Institute (RTRI, Japan) and Lueå Railway Research Centre (JVTC, Sweden).

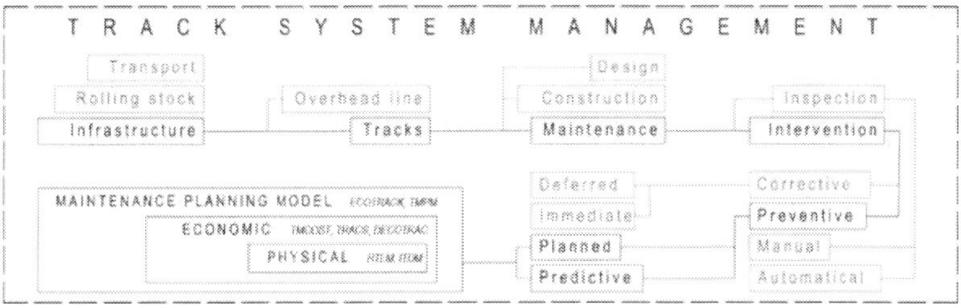

Figure 1. Maintenance models within rail infrastructure management process.

TMCOST—Track Maintenance Cost Model is a result of cooperation between TTCI and Massachusetts Institute of Technology (MIT). It consists of a series of separate submodels for rail, sleeper and ballast degradation assessment as a function of traffic load. It calculates maintenance costs by assessing the lifetime of individual track component and maintenance costs needed for achieving initial track quality [13].

RTLM—Railroad Track Life Cycle Model is developed by the TTCI. It consists of a series of separate submodels for track vertical geometry degradation based on type, thickness and condition of ballast material, sleeper type and spacing, rail type, subgrade bearing capacity, designed vertical geometry, axle load and traffic volume [10].

TRACS—Total Right-of-Way Analysis and Costing System is a result of cooperation between TTCI and MIT. It was created as a TMCOST model upgrade. It estimates track maintenance and renewal costs as a function of the track route design geometry, track components, and traffic volume and type [14].

ITDM—Integrated Track Degradation Model is a result of cooperation between QR and Queensland University of Technology (QUT). It estimates track degradation as a function of axle load and rolling stock speed. It consists of four submodels for rail, sleeper, ballast and subgrade degradation. It predicts both individual subsystem degradation and whole track system degradation [15].

ECOTRACK—ECOnomical TRACK is a result of cooperation between ERRI and International Union of Railways (UIC). This is a simple model that estimates tracks exploitation costs without detailed modelling of transport and track construction characteristics. It uses the measured track geometry values integrated in historical database. It predicts the geometry degradation rate by linear regression [5, 6, 10].

TMPM—Track Maintenance Planning Model is developed by QUT by integrating ITDM model into predictive maintenance interval and cost model [16].

DECOTRACK—Degradation Cost Of Track is a result of cooperation between Swedish Rail Administration (Banverket) and JVTC. It predicts rail degradation during exploitation in the form of fatigue and wear. Its input parameters are axle load, annual transport capacity in tonnes, speed, type and state of vehicle maintenance. The model results in anticipated rails lifetime and maintenance costs [11].

2.3. The possibility of developed models application on tram systems

The knowledge on degradation analysis and modelling of tram tracks is still rudimentary [17, 18]. The adoption of best practices in the establishment of a modern tram track maintenance system is certainly preferable than starting the process from scratch. However, the knowledge gained about conventional ballasted intercity rail system degradation cannot be directly applied to the narrow gauge tracks in urban areas. This is due to significant differences in the design requirements and exploitation conditions that are defined as key factors of any rail system degradation. These differences are related to the tram tracks location, design geometry, construction, vehicles and traffic organization [2].

The proximity between the tram tracks and the surrounding facilities and requirements for the rational use of city traffic areas during the tram route design demands narrower track gauges and tighter horizontal curves. For this reason, the tram vehicles are of smaller dimensions and weight. Axle loads on conventional railways are ranging from 16 to 25 tons per axle, whereas the axle load on urban tram tracks are generally lower than 13 tons per axle [19]. Also, the construction of trams undercarriage is different from conventional trains, and this difference is particularly pronounced in the case of modern low-floor trams. The differences are also reflected in the tram traffic flow characteristics given the movement priority and speed of the rolling stock. Due to a lack of space in densely built urban centres, trams often have to share the lanes with road vehicles and therefore adjust (primarily reduce) their speed. For the same reason, tram tracks superstructure is usually built on continuously reinforced concrete slabs, with grooved rails enclosed in pavement construction. The use of such rails also means that tram wheels flange is narrower than one on train wheel.

The review of the current practice has shown us that conclusions about the behaviour of conventional rail tracks during exploitation can only be used as a basis for further research of tram tracks degradation. In addition, significant differences between numerous tram systems construction, traffic conditions and monitoring procedures [2, 20, 21] complicate determining universal rules of narrow gauge tram track degradation. This indicates the need for research activities that would result in creation of a new approach to monitoring narrow gauge tram tracks behaviour during exploitation and the development of a mathematical model of tracks degradation, taking into account joint effect of track design and exploitation conditions.

3. Creation of physical tram track gauge degradation model

A key element of the coordinated maintenance planning based on the track degradation modelling is up to date, digital, concise and widely available historical database on the tracks. It contains comprehensive information on tracks design and construction elements, exploitation conditions and monitoring and maintenance history (conducted inspections, activities and maintenance costs).

Although tracks monitoring and modelling are recognized as very important activities, many tram networks managements face similar problems when attempting to create such database. This is because most European cities have retained the traditional tram systems that date back to before the First World War. Nowadays, they face the problem of documenting their infrastructure in digital databases. In addition, the actual knowledge on the tracks condition is limited to a small number of managements' employees. Also, most of these employees do not have any tools for collecting, systematizing and integrating data in the historical database on which to conduct the assessment of the maintenance needs. Fortunately, cities that have recently upgraded their tram systems into modern networks based on the technologies and practices of urban light rail now have a large amount of precise digital documentation necessary for creation of historical databases [22, 23]. One such system is the tram system in the city of Zagreb, with about 80 km of tram tracks that were built and reconstructed during the past two decades.

Today, regular tram traffic in Zagreb is organized in 15 lines served daily by 178 tram vehicles operating on 116.3 km of 1000 mm gauge tracks. About half of the operational length of the tracks is placed in a separate tram corridor, whereas the rest share their corridor with road traffic (**Figure 2**) [24]. Two types of steel grades for grooved Ri-60 rails are used on Zagreb tram tracks: steel grade R200 rails at tangential and curved tracks with radius R≥200 m and wear-resistant steel grade R260 rails at curved tracks with radius R<200 m. Rails are discreetly laid on the levelling layers, made out of micro synthetic concrete, which are built on reinforced concrete slab. The distance between levelling layers is one meter, and rails are fixed to them by elastic fastening systems.

Exploitation conditions on Zagreb tram tracks are very harsh: individual sections have a traffic volume of up to 15 million gross tonnes (MGT) per year, with vehicle passing frequency of <90 s, and loads of more than 3.5 tonnes per wheel [25]. This high-capacity network presents an optimal testing ground for exploration of the possibilities to introduce the predictive maintenance system on narrow gauge tram tracks through the development of a track degradation model, based on the principles established on conventional rail track structures.

One of the main factors ensuring the tram traffic safety and ride comfort is maintaining high-quality track geometry. The required tram traffic safety refers to prevention of tram derailment. The required ride comfort refers to limiting the amount of lateral movement of trams in motion.

According to Ref. [26], tram track geometry is defined by track gauge (G) and cross level (h). *Track gauge* is defined as the spacing of the rails measured between their inner faces, 9 mm below the rail head. On newly built, reconstructed and repaired tangential tram tracks, gauge must be 1000 (+3, −2) mm. Maximum permissible deviation from this basic gauge in exploitation (ΔG) ranges from −2 to +25 mm. *Cross level* is defined as the difference in elevation (height) between the inner and outer rail in horizontal curve. Its value is determined by designer, depending on the curve radius and the design speed. It should be noted that the cross level is predicted only for tram tracks in separate corridors. If the trams and road vehicles share the same traffic area, cross level will not be implemented as it is incompatible with the pavement surface. For this reason, in the study of tram tracks geometry degradation, we will concentrate only on gauge degradation modelling.

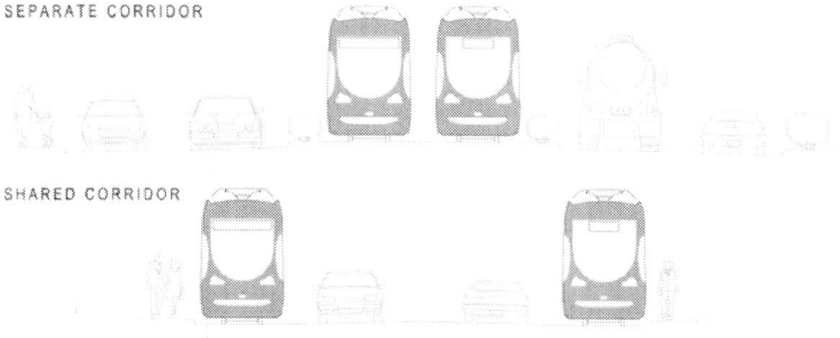

Figure 2. Types of Zagreb tram corridors—separate and shared.

When viewed in the ground plan, the trajectory of the moving rail vehicles centre of gravity has the shape of a sinusoid [27]. This hunting motion is a consequence of differences in the spacing between wheel flanges and the track gauge. This kind of motion gets more pronounced as the difference, that is, track gauge increases, primarily due to rail wear. Rail wear is a consequence of wheel rolling and sliding contact abrasion, and it is manifested as loss or movement of material in the contact area of rail head (**Figure 3**).

Besides compromising requested ride quality and smoothness, rail wear, that is track gauge increase, causes additional dynamic loads on track, its faster degradation and higher maintenance costs. Therefore, it is not surprising that measuring and modelling of gauge degradation are the most common forms of monitoring rail infrastructure within the planned maintenance system.

3.1. Choosing the modelling approach

Considering the *modelling approach*, applied type of statistical data processing (inferential or descriptive) and therefore needed input data and analysis complexity, mathematical models of track geometry degradation can be divided into probabilistic, mechanistic, empirical and mechanistic-empirical models (**Figure 4**) [5, 6].

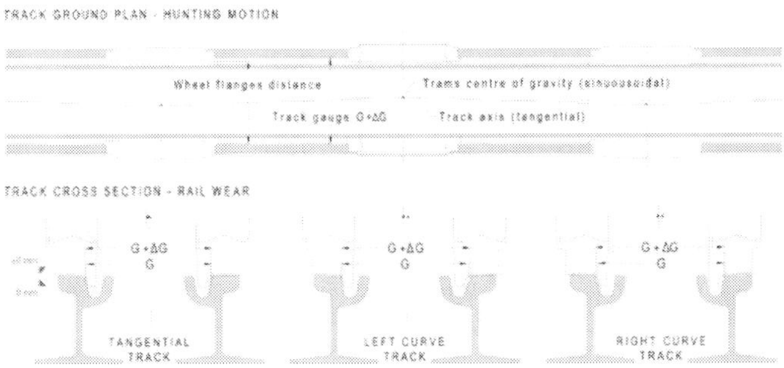

Figure 3. Hunting motion and tram track gauge increase due to rail head wear.

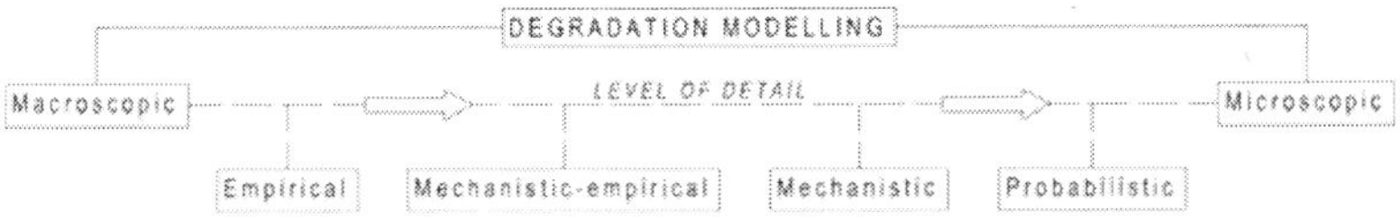

Figure 4. Input data and analysis complexity—degradation models hierarchy.

Probabilistic model uses the inferential statistics techniques to define probability that the quality of particular track section in one time period will be the same in the next period [5, 28–34]. The application of such degradation models showed that exclusive use of statistically determined probability of track degradation, instead of using actual measured data, does not provide a complete and accurate assessment of the track quality [23].

Mechanistic model simulates track degradation using well-established theoretical principles of track components behaviour during exploitation [35]. This so-called engineering approach uses the inferential statistics techniques to predict the degradation rate of the track, of certain initial quality, on the basis of known exploitation conditions. These models can be very sophisticated and often require the use of complex computer algorithms and long-term calculations. However, they can simulate the tracks behaviour in different conditions, even before its exploitation, and help to define tracks most important properties needed for improving its exploitation behaviour [36–39].

Empirical models are based on track quality data obtained experimentally or by field observations. They define degradation rate of individual track sections (so-called track degradation evolution [12, 36]) using descriptive statistics methods. They put into correlation the data on systematic track monitoring and exploitation period defined through track exploitation time or intensity in million gross tonnes [4, 40, 41]. Model results may significantly vary depending on the data used for their development and therefore should not be used to predict the track degradation [36]. They are mostly used for a better understanding of the observed degradation processes, as a first step of creating or for validating other types of models [42, 43].

Mechanistic-empirical model is based on a combination of mechanistic and empirical modelling procedures. This is considered the most effective track degradation modelling approach [44]. Creating such model requires track segmentation, that is separation of linear rail infrastructure on segments with homogeneous characteristics of degradation influential factors, monitoring and maintenance history. By summarizing the information obtained through track monitoring on each observed segment, it is possible to predict the need for maintenance of the whole network for a certain exploitation period [4–6]. This model determines the track degradation rate by statistical regression analysis using least square method [6, 45]. It is usually carried out over the average values of the track quality parameters, calculated from the values measured in individual points (chainages) of track section [46]. Regression model defines the degradation rate of the dependent variable—the observed track quality parameter, as a function of independent variable—the exploitation period expressed as time or exploitation intensity [4]. This allows defining critical maintenance limits for the observed geometry parameter or track element in the investigated sections [47]. In addition, there is a recent trend of track degradation modelling using artificial neural networks. Its aim is to achieve a rational insight into the track exploitation behaviour without its monitoring [48].

Through review of the development, preparation and characteristics of railway tracks geometry degradation models, two main objections of their end users were identified. Models are either too general and do not take into account the specific rail traffic and track characteristics, or too specific, that is they require a large amount of input data and interpretation of their results requires a high level of insight into the degradation problem. That is why today most researchers aim to develop as simple as possible degradation models that are convenient for the implementation and interpretation.

After modelling approaches, characteristics and procedures analysis, it was concluded that modelling of tram track gauge degradation during exploitation, due to the availability and format of the data on the Zagreb tram tracks, should be carried out using mechanistic-empirical approach.

3.2. Gauge degradation influential factors identification

Track geometry degradation in the form of gauge increase caused by rail wear is the subject of numerous research studies conducted to accurately determine the mechanisms of its development. Studies have shown that this process is very complex and depends on a number of influential factors. They can be divided into three basic groups: *traffic*, *construction* and *design geometry* influential factors.

Gauge degradation is primarily a result of the dynamic loading from vehicles running on tracks. These forces occur because of the irregularities in wheel-rail contact surface and track geometry. The rate of degradation is proportional to the exploitation intensity and vehicles speed. With regard to construction factors, studies have shown that the increase in rail steel hardness slows down the process of gauge degradation. Horizontal track geometry design elements also have a major effect on the degradation process, in particular the track curvature. Gauge degrades faster in horizontal curves than in tangential track sections, and the degradation rate is proportional to tracks curvature [21, 44, 49–52].

For the purpose of tram track gauge degradation modelling, beside exploitation intensity, the following six design, construction and traffic characteristics were defined as influential: track curvature, rail quality, that is steel hardness, rail fastening system stiffness, paving system type, estimated tram operation speed due to the corridor type and due to arrangement of stops and crossings. The aim of the model creation is to determine the individual contribution of influential factor on tram track gauge degradation rate.

3.3. Segmented database creation

For the purpose of this investigation, a review of the available construction and supervision documentation made for the (re)construction of tram tracks in Zagreb in the period 1997–2004 was performed. It resulted in identification of 11 sections of the network (more than 26 km of tram tracks) suitable for model creation.

Reviewed documentation included the results of control track gauge measurements conducted just before the commissioning of the (re)constructed track. Gauge measurements were carried out in the tram travelling direction, above every discrete rail levelling layer, that is at 1-m interval. Measured gauge values were denoted with corresponding track chainage values (geographic locations of gauge measurement cross sections along the track) defined in digital georeferenced track design blueprints. During gauge measurements, characteristic track superstructure element sections chainages were observed and recorded. Also, locations of specific track cross sections (locations of rail welds, station platforms and road crossings) were marked.

In order to determine gauge degradation due to track exploitation, continuous gauge measurements in the same track cross sections were repeated during spring months of 2011 and 2013.

For these tram tracks, described by more than 26×10^3 georeferenced cross sections, a comprehensive historical database of geographically synchronized modelling input data was made. In addition to the gauge values, to each section, an observed traffic, construction and design geometry characteristics and the value of exploitation intensity were assigned.

Calculation of exploitation intensity was conducted by integrating data from the available Zagreb Municipality Transit System—ZET Ltd. internal documents. These documents include information about an hourly frequency of vehicles on a single tram line, the daily number of vehicles of a certain type on each tram line, tram lines network maps and approximations of vehicles capacity utilization. Cumulative exploitation intensity was defined as the product of total number of exploitation days (defined by track gauge measurements dates on observed track segments) and daily gross mass of trams with passengers (in MGT).

Created database, which contains more than 34×10^4 quantitative and qualitative data on tram tracks, was then divided into 425 segments, that is linear track gauge data sets with homogeneous characteristics of gauge degradation influential factors. A small portion of segmented database is shown in **Figure 5**.

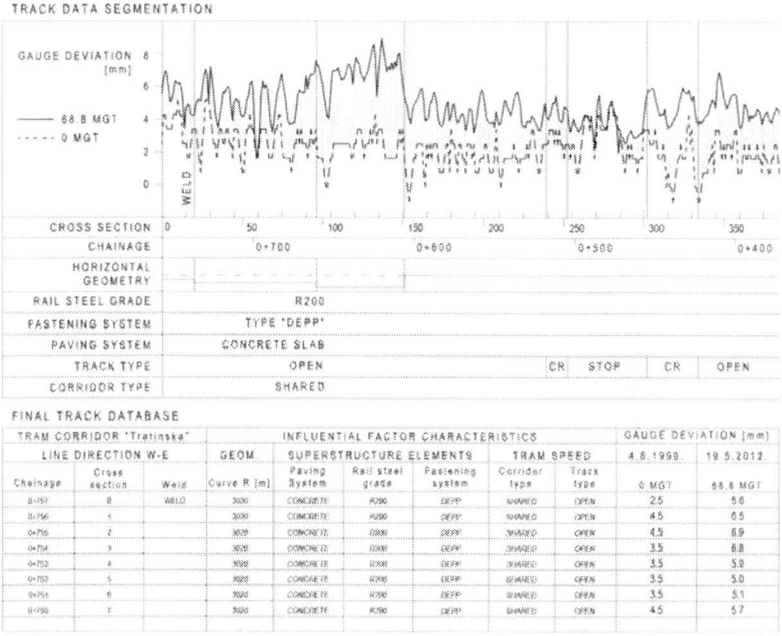

Figure 5. Portion of segmented track database.

3.4. Gauge data filtering and compression

On the basis of the gauge values measured after different exploitation periods, deviation values from the prescribed gauge of 1000 mm were calculated in each measurement cross section. In order to minimize possible measurement and/or geographical data synchronization errors, the following steps of track gauge deviation data filtering were carried out.

Mean gauge deviations differences and standard deviations were calculated on each of the identified 425 segments, in the corresponding exploitation period. Outliers were identified along each segment as individual gauge deviation differences which deviate from the mean of the segment for more than triple of the standard deviation value. These outliers were substituted with the first larger or smaller non-outlier gauge deviation difference value within the same segment. Segments that showed high data variability were excluded from further analysis. In general, excluded segments were ones in crossings, in tram stops where there was an occurrence of rail plastic flow and less than 30 m in length.

After this initial filtration, data compression and filtration were continued along each segment, track and then section. In this way, 35 representative and, according to observed gauge degradation influential factors (all but exploitation intensity—it will be used as independent variable in regression analysis), characteristic types of tram tracks were identified.

3.5. Regression analysis

It is a well-known fact that track gauge degradation process is more pronounced in a short period immediately after track construction or renewal, that is new rail installation or old rail reprofiling. This period of initial severe rail abrasion, during which the rail profile adapts to the shape of wheels flange, is followed by a period of more gradual rail degradation. In this gradual increase, which directly affects the track gauge, *linear trend* was observed [9]. This means that gauge degradation can be quantified in a linear mathematical equation, which describes the relationship between gauge deviation and track exploitation period, taking into account degradation influential factors.

By regression analysis of the relationship between compressed gauge deviation difference values and the track section exploitation intensity, linear function was defined for each of the observed 35 types of tracks. Model slopes, that is regression coefficients, define the modelled rate of gauge degradation during exploitation for each characteristic type of track. Example of linear gauge degradation models for four types of tram tracks is shown in **Figure 6**.

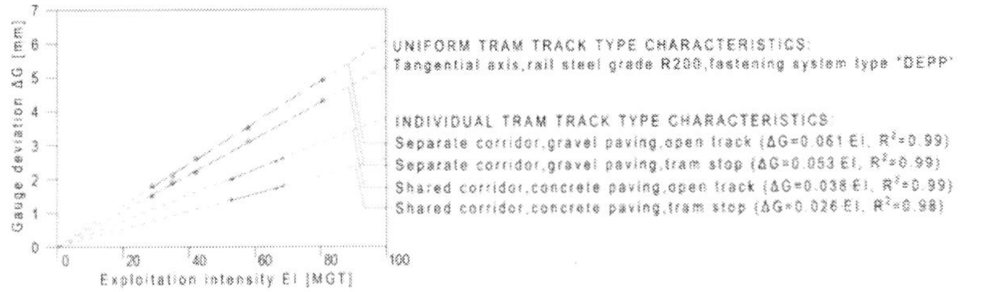

Figure 6. Example of linear gauge degradation models for four types of tram tracks.

Analysis of the models representativeness showed that there is a strong link between the observed variables in all 35 cases ($0.74<R^2<0.99$) and that average model residuals (±0.3 mm) in regard to the achievable accuracy of measurement (±0.3 mm), as well as their variability (<20%), are satisfactory.

4. Tram track gauge degradation modelling results

Comparison of models regression coefficients showed that track gauge degradation rate is as follows:

- smallest at tangential sections of tram tracks on stops in shared road corridor, built with head hardened, elastically fastened rails enclosed with concrete paving slabs and
- largest at separated open track sections, in horizontal curves with radii less than 300 m, built with wear-resistant rails enclosed with gravel.

Further analysis of modelled regression coefficients relationships showed that the observed track gauge degradation influential factors, by the amount of their average impact, may be ranked as follows:

- Tram speed dependent on the tram corridor type has the greatest impact on the gauge degradation rate. In the case of the Zagreb tram network, the ability to develop higher travelling speeds is achievable exclusively in separate tram corridors that extend through avenues and major streets central belts. The tram speed on shared road corridors is limited by the behaviour of other road users, and typically shorter distances between tram stops and signalized intersections. In general, higher tram speed causes larger dynamic forces on the track, that is more prominent rail damage and track geometry degradation during exploitation.
- The second largest influence on the gauge degradation rate has track horizontal curvature. The results of this research showed that, of course, higher curvature increases gauge degradation rate, but also that the influence of track curvature on outer grooved rail wear is neglectable in curves with radius larger than 1200 m. Additionally, it was observed that the distribution of gauge deviation values along the horizontal curve depends on curve radius and length. In curves with radius less than 300 m and longer than 50 m, maximum rail wear occurs at the end of the curve (given the tram travelling direction). Otherwise, gauge degradation is more prominent in the vertex zone of the curve.
- The third gauge degradation influential factor is track superstructure elasticity. In average, gauge degradation rate is higher when tram tracks are built using stiffer rail fastening system. More detailed elaboration of this tram track gauge degradation influential factor is given in Ref. [53].
- The track quality defined by its rail tensile strength is the next tram tracks gauge degradation influential factor in order of relevancy. As expected, the use of higher grade steel rails can slow down the rate of rail head wear during exploitation. However, the results of this research showed that, in order to reduce rail wear in curves, head hardened wear-resistant rails should be used.
- Fifth in-line gauge degradation influential factor is corridor type according to the arrangement of tram stops. Research has shown that the positive effect of dynamic forces reduction by reducing the tram speed to zero along the tram stops is annulated by the occurrence of additional dynamic effects caused by trams decelerating and accelerating.
- Influence of paving system used for enclosing the tram tracks (by either gravel or concrete slabs) on gauge degradation rate is little to none, if we exclude the effect of differences between achievable tram speed on different track corridor types (shared or separate).

5. Concluding remarks and recommendations for future work

Developed gauge degradation models represent the first small step towards establishing a preventive maintenance system on Zagreb tram network. For now, they can provide only an

http://dx.doi.org/10.5772/65086

insight in (by design, constructive and traffic characteristics specific) track sections degradation behaviour. Although the models are of satisfactory representativeness, the overall process of model creation pointed out the following challenges.

The research was limited by the availability and form of the input data about tram tracks required for the creation of database over which the modelling would be carried out. These data were collected and stored over the years by various stakeholders for numerous reasons, other than modelling gauge degradation. It is our recommendation, prior to any future extensive trams network (re)constructions, to establish procedures for detailed recording of these characteristics and their integration into a single database. Also, keeping track of exploitation parameters of network's individual track sections, compared to the current practice of keeping records of exploitation parameters of tram lines, would significantly simplify and therefore accelerate the process of exploitation intensity calculation.

After final step of gauge deviation data compression and filtering, tram track segments along the road crossings were excluded from further analysis. Large variability of empirical data along these segments suggests that road vehicles at crossings definitely affect the gauge degradation. However, the comparison of the calculated average values of gauge deviations along these segments showed no regularity. It was concluded that for the establishment of track gauge degradation model on these segments, it is necessary to include data on the intensity of road traffic which they are exposed to.

To conclude, for small urban rail networks such as tram network in Zagreb, the use of deterministic mechanistic-empirical approach and statistical (regression) analysis in track degradation modelling has provided specific and useful new insights into the tracks behaviour during operation. Such modelling approach was adopted rather than probabilistic, multiparameter one. It is our belief that creation of simple and easy to use model, rather than comprehensive simulation tool, would initiate sooner modernization of tram track maintenance procedures and regulations. However, for gradual increase in portions of network that could serve as a platform for further research and creation of predictive tram tracks degradation models, a closer cooperation between tram network managements design, construction, maintenance and transport organization divisions is needed.

Author details

Maja Ahac* and Stjepan Lakušić

*Address all correspondence to: mahac@grad.hr

Department for Transportation Engineering, Faculty of Civil Engineering, University of Zagreb, Zagreb, Croatia

References

[1] Life Cycle Cost Optimisation. A UITP (International Association of Public Transport) information sheet. Metropolitan Railways Committee, Electrical Installations & Safety Systems Subcommittee Core Brief. Brusseels; 2009. Available from: https://go.itdp.org/download/attachments/42598847/19)%20UITP%20Core%20Brief%20-%20Life%20Cycle%20Cost%20Optimization%20(EU%202009).pdf?api=v2

[2] Visual Inspection and Maintenance - Proposal for European standard for track inspection and maintenance. URBAN TRACK Project. 2009. Available from: http://www.urbantrack.eu/images/site/deliverables/D0204_FCP_M36.pdf

[3] Belak S. Terotechnology. 1st ed. Šibenik: Manualia Collegii Studiorum Sibenicensium; 2005. 225 p.

[4] Khouy I A. Optimization of Track Geometry Maintenance - A Study of Track Geometry Degradation to Specify Optimal Inspection Intervals [thesis]. Luleå: Luleå University of Technology; 2011. 98 p. Available from: http://pure.ltu.se/portal/files/36779920/Iman_Arasteh_khouy_Lic2011.pdf

[5] Andersson M. Strategic Planning of Track Maintenance - State of the Art. Research Report TRITA-INFRA 02-035. Borlänge: KTH Royal Institute of Technology; 2002. 59 p.

[6] Jovanovic S. Railway track quality assessment and related decision making. In: International Conference on Systems, Man and Cybernetics (Vol 6); 10–13 October; The Hague. New York: IEEE; 2004. p. 5038–5043. DOI: 10.1109/ICSMC.2004.1400992

[7] Lakušić S, Lazarević D, Dvornik J. Dynamic analysis of the tramcar & track interaction. GRAĐEVINAR. 2006;58(4):284-288.

[8] Karttunen K. Mechanical track deterioration due to lateral geometry irregularities [thesis]. Goeoteborg: Chalmers University of Technology; 2012. 32 p. Available from: http://publications.lib.chalmers.se/records/fulltext/168543/168543.pdf

[9] Esveld C. Modern Railway Track. 2nd ed. Delft: MRT-Productions; 2001. 623 p.

[10] Roberts C M. A prototype decision support system for effective track maintenance and renewal – Appendix B: The development of track deterioration models [thesis]. Birmingham: University of Birmingham; 2001. 19 p.

[11] Larsson D. A study of the track degradation process related to changes in railway traffic [thesis]. Luleå: Luleå University of Technology; 2004. 82 p. Available from: http://epubl.ltu.se/1402-1757/2004/48/

[12] Fortunato E, Pinelo A, Lobo Costa J, Gonçalves D, Pratas A. Some results of performance indicators of a railway track obtained during its renewal process. In: 4th International SIIV Congress; 12-14 September; Palermo. 2007.

[13] Reddy V. Development of an integrated model for assessment of operational risks in rail track [thesis]. Brisbane: Queensland University of Technology; 2007. 279 p. Available from: http://eprints.qut.edu.au/16488/1/Venkatarami_Reddy_Thesis.pdf

[14] Shughart L A. Reconstructing a railroad engineering economics model: responding to management needs [thesis]. Cambridge: Massachusetts Institute of Technology; 1989. 330 p. Available from: dspace.mit.edu/bitstream/handle/1721.1/33247/21793122-MIT.pdf?sequence=2

[15] Zhang Y-J, Murray M, Ferreira L. Modelling Rail Track Performance: An Integrated Approach. Transport. 2000;141(4):187-194. DOI: 10.1680/itran.2000.32786

[16] Simson S, Ferreira L, Murray M. Rail track maintenance planning: an assessment model. Transportation Research Record: Journal of the Transportation Research Board. 2000;1713(1713):29–35. DOI: 10.3141/1713-05

[17] Ahac M. Mechanistic-empirical model for tram track gauge deterioration during exploitation [thesis]. Zagreb: University of Zagreb; 2013. 242 p.

[18] Yousefikia M, Moridpour S, Setunge S, Mazloumi E. Modeling degradation of tracks for maintenance planning on a tram line. Journal of Traffic and Logistics Engineering. 2014;2(2):86-91. DOI: 10.12720/jtle.2.2.86-91

[19] Law on the Railway Transport Safety, Official Journal of the Republic of Croatia - Narodne novine. 40/07 and 61/11. Available from: http://narodne-novine.nn.hr/clanci/sluzbeni/2007_04_40_1343.html

[20] Track Design Handbook for Light Rail Transit - TCRP Report 57. Washington D.C.: Transportation Research Board. National Academy Press; 2000. 84 p. Available from: http://www.tcrponline.org/PDFDocuments/tcrp_rpt_57-a.pdf

[21] Preventive maintenance of embedded tram tracks – rail wear in curves and special trackwork for trams. URBAN TRACK Project. 2008. Available from: http://www.urban-track.eu/images/site/deliverables/d0208_stib_m24.pdf

[22] Andersson M, Murray M, Ferreira L, Lake N. Collection and Use of railway track performance and maintenance Data. In: Proceedings of the Conference on Railway Engineering (CORE); 20-23 June; Darwin. 2004.

[23] Madejski J. Light rail, tram track and turnout geometry measurement and diagnostic tools. In: Brebbia C A, Wadhwa L, editors. Urban Transport XI. - Urban Transport and the Environment in the 21st Century. Southampton: WIT Press; 2005. p. 185–196. DOI: 10.2495/UT050191

[24] Zagreb holding Ltd. - Zagreb Municipality Transit System ZET [Internet]. http://www.zet.hr. [Accessed: 2013-06-03]

[25] Lakušić S, Haladin I, Ahac M. The effect of rail fastening system modifications on tram traffic noise and vibration. Shock and Vibration. 2016;2016:4671302. DOI: 10.1155/2016/4671302

[26] Regulation on track and track facilities maintenance. Zagreb Municipality Transit System ZET. Zagreb; 1997.

[27] Abood K H A, Khan R A. Hunting phenomenon study of railway conventional truck on tangent tracks due to change in rail wheel geometry. Journal of Engineering Science and Technology. 2011;6(2):146-160.

[28] Perrin G, Soize C, Duhamel D, Funfschilling C. Track irregularities stochastic modelling. Probabilistic Engineering Mechanics. 2013;34:123-130. DOI: 10.1016/j.probengmech.2013.08.006

[29] Perrin G, Soize C, Duhamel D, Funfschilling C. Modeling the track geometry variability. In: 10th World Congress on Computational Mechanics (WCCM); 8-13 July; São Paulo. 2012. DOI: 10.5151/meceng-wccm2012-16655

[30] Oyama T, Miwa M. Mathematical modeling analyses for obtaining an optimal railway track maintenance schedule. Japan Journal of Industrial and Applied Mathematics. 2006;23(2):207–224. DOI: 10.1007/BF03167551

[31] Andrade A R, Teixeira P F. A Bayesian model for rail track geometry degradation: a decisive step towards the assessment of uncertainty in rail track life-cycle. In: 12th World Conference on Transport Research (WCTR); 11-15 July; Lisbon. WCTRS; 2010.

[32] Zakeri J-A, Shahriari S. Developing a deterioration probabilistic model for rail wear. International Journal of Traffic and Transportation Engineering. 2012;1(2):13-18. DOI: 10.5923/j.ijtte.20120102.02

[33] Quiroga L, Schnieder E. Modelling of high speed railroad geometry ageing as a discrete-continuous process. In: Stochastic Modeling Techniques and Data Analysis International Conference (SMTDA 2010); 8–11 June; Chania, Crete. 2010. p. 655-842.

[34] Shafahi Y, Hakhamaneshi R. Application of a maintenance management model for Iranian railways based on the Markov chain and probabilistic dynamic programming. Scientia Iranica. 2009;16(1):87-97.

[35] Crawford S, Murray M, Powell J. Development of a mechanistic model for the determination of track modulus. In: 7th International Heavy Haul Conference; 10-14 June; Brisbane. 2001.

[36] TRA - Railway Tie Association. Comprehensive rail and track related research [Internet]. Available from: http://www.rta.org/comprehensive-rail-and-track-related-research [Accessed: 2012-05-18]

[37] Heo G Y. Condition monitoring using empirical models: technical review and prospects for nuclear applications. Nuclear Engineering and Technology. 2008;40(1):49-68. DOI: 10.5516/NET.2008.40.1.049

[38] Zhang Y-J, Carr G, Fateh M. Microscopic modeling of gage widening. Transportation Research Record: Journal of the Transportation Research Board. 2002;1785(1):64-70. DOI: 10.3141/1785-09

[39] Orvnäs A. Simulation of Rail Wear on the Swedish Light Rail Line Tvärbanan [thesis]. Stockholm: KTH - Royal Institute of Technology; 2005. 58 p. Available from: http://www.diva-portal.org/smash/get/diva2:371655/fulltext01.pdf

[40] Lyngby N. Railway track degradation: shape and influencing factors. International Journal of Performability Engineering. 2009;5(2):177-186.

[41] Zhou R R, Serban N, Gebraeel N. Degradation modelling applied to residual lifetime prediction using functional data analysis. The Analysis of Applied Statistics. 2011;5(2B): 1586-1610. DOI: 10.1214/10-AOAS448

[42] Ishida M, Takikawa M, Jin Y. Gauge face wear caused with vehicle/track interaction. In: World Conference on Railway Research 2001; 25-29 November; Koeln. WCRR; 2001.

[43] Whole Life Rail Model Application and Development: Six Month Progress Report. AEA Technology Rail Report. RSSB Railway Safety No. AEATR-TEaM-2003-231; 2003. 149 p.

[44] Sadeghi J, Askarinejad H. Influences of track structure, geometry and traffic parameters on railway deterioration. International Journal of Engineering – Transactions B: Applications. 2007;20(3):291-300.

[45] Premathilaka A, Costello S, Dunn R. Development of a deterministic rail-wear prediction model. Road and Transport Research. 2010;19(1):40-50.

[46] Pedanekar N R, Venkoba Rao B, Kapur P C. Profile-based prediction: an enhanced methodology for prediction of track degradation. In: AREMA 2003 Annual Conference; 5-8 October; Chicago. American Railway Engineering and Maintenance-of-Way Association; 2003.

[47] Kawaguchi A, Miwa M, Terada K. Actual data analysis of alignment irregularity growth and its prediction mode. Quarterly Report of RTRI. 2005;46(4):262-268. DOI: 10.2219/rtriqr.46.26

[48] Sadeghi J, Askarinejad H. Application of neural networks in evaluation of railway track quality condition. Journal of Mechanical Science and Technology. 2012;26(1):113-122. DOI: 10.1007/s12206-011-1016-5

[49] Olofsson U, Lewis R. Chapter 5: Tribology of the wheel – rail contact. In: Iwnicki S, editor. Handbook of Railway Vehicle Dynamics. Boca Raton: CRC Press – Taylor and Francis Group; 2006. p. 121-141.

[50] Sadeghi J, Akbari B. Field investigation on effects of railway track geometric parameters on rail wear. Journal of Zhejiang University – Science A: Applied Physics & Engineering. 2006;7(11):1846-1855. DOI: 10.1631/jzus.2006.A1846

[51] Faiz R B, Singh S. Information analysis of rail track for predictive maintenance. WSEAS Transactions on Computers. 2009;8(7):1123-1133.

[52] Povilaitiene I, Kamaitis Z, Podagelis I. Influence of gauge width on rail side wear on track curves. Journal of Civil Engineering and Management. 2006;12(3):255-260. DOI: 10.1080/13923730.2006.9636400

[53] Ahac M, Lakušić S. Tram track maintenance-planning by gauge degradation modelling. Transport. 2015;30(4):430-436. DOI: 10.3846/16484142.2015.1116464

Chapter 2

Design of Energy Management System of a PEMFC–Battery–Supercapacitor Hybrid Tramway

Qi Li and Weirong Chen

Additional information is available at the end of the chapter

http://dx.doi.org/10.5772/64696

Abstract

In this chapter, a hybrid power system that consists of multiple proton exchange membrane fuel cell (PEMFC) systems, batteries, and supercapacitors (SCs) is developed for a hybrid tramway. Three energy management strategies that included a fuzzy logic control (FLC), an equivalent consumption minimization strategy (ECMS), and a state machine strategy based on droop control (SMS-DC) are utilized to coordinate multiple power sources, avoid the transients and rapid changes of power demand, and achieve high efficiency without degrading the mechanism performance for an energy management system of hybrid tramway. A hybrid system model of tramway is established with commercially available devices, and then the different energy management strategies are evaluated with a real driving cycle of tramway from Turkey. The results compared with these strategies demonstrate that the higher average efficiencies of the tramway, the lower tramway-equivalent hydrogen consumptions, and more efficient use of the batteries and SCs energy are achieved by the SMS-DC. Therefore, the appropriate energy management system for high-power hybrid tramway will improve the hydrogen consumptions of overall hybrid system and the efficiencies of each power source.

Keywords: proton exchange membrane fuel cell, hybrid tramway, energy management system, a state machine strategy, fuzzy logic control, equivalent consumption minimization strategy

1. Introduction

Nowadays, environmental issues and energy crisis relating to oil supply, pollution, and green house effects justify the need for developing of new technologies for transportation. Fuel cells as a promising technology, which provide electrical power with high efficiency, less noise, and near zero emissions compared with conventional internal combustion engines, are expected to become a viable solution for transportation applications [1, 2]. Due to its lower operating temperature and higher efficiency, a proton exchange membrane fuel cell (PEMFC) is considered as one of the most prime candidates for the vehicular applications [3–5].

In recent years, several research efforts have been carried out to develop the locomotives and the tramways powered by the PEMFC in order to encourage hydrogen economy development and reduce dependence on fossil fuels [6–12]. The locomotives and the tramways powered by the PEMFC and the energy storage system (ESS) contrasted with the diesel-electric and catenary-electric types have many advantages, which show great extensive application potential. The PEMFC exhibits good power capability during steady-state operation, but its response limitation for transient peak power demand has restrained the PEMFC from being used in large-scale and high-power transportation applications, and the lifetime of the PEMFC may be dramatically impacted by the rapid power demand variations. Due to the unidirectional power flow characteristics of the PEMFC, the energy from regenerative braking of a vehicle cannot be handled by the PEMFC. Hence, hybridizing the PEMFC with the ESS, such as battery, supercapacitor (SC), or a combination of both, can meet the total power demand and reduce the effect of drawback from the powertrain driven by stand-alone PEMFC. The hybrid tramway composed of multiple motor and trailer units, such as the type of two motor units and one trailer unit, is driven by multiple PEMFC and ESS based on the requirement of the power level and the mounting space. In order to fulfill the power balance between the load power demand and the power sources, the energy flows are distributed by the energy management system, which determines the power generation split at each sampling time between the PEMFC and the ESS, such as battery and SC.

Recently, the development of energy management system has become a topic of interest for researchers. Jia et al. [4] have presented the electrical characteristic of a hybrid power supply system combining PEMFC and SC and have investigated on the platform of an electric bicycle to improve the system efficiency. Xu et al. [5] have proposed a multimode real-time control strategy based on three typical processes of the fuel cell system for a fuel cell electric vehicle, taking the fuel economy and system durability into consideration. Li et al. [6, 7] have proposed a power sharing strategy for an energy management system of hybrid tramway and also have presented a state machine strategy based on droop control to coordinate multiple power sources of hybrid tramway. García et al. [10] have presented an equivalent consumption minimization strategy for the energy management system of the fuel cell/battery/SC tramway. Torreglosa et al. [11] have optimized the tramway hydrogen consumption based on fuel cell/battery hybrid system by the equivalent consumption minimization strategy. Fernandez et al. [12] have utilized a state machine control to satisfy the power demand from the real driving cycle for the fuel cell/battery hybrid tramway. Thounthong et al. [13] have fulfilled the comparisons between linear proportional integral and nonlinear flatness-based controllers for DC link stabilization. Eren et al. [14] have proposed a fuzzy logic supervisory controller-based power management strategy that secures the power balance in hybrid structure, enhances the FC performance, and minimizes the power losses for an FC/UC hybrid vehicular power system.

In this chapter, a hybrid propulsion system including two PEMFC systems, two batteries, and two SCs is developed for the tramway with two motor units and one trailer unit. The PEMFC systems act as main power source of tramway, and the Li-ion batteries and the SCs are utilized as the ESS to supplement the PEMFC output power during tramway acceleration and cruise and are also utilized for energy recovery during braking. A hybrid system model of tramway is established with commercially available devices. Furthermore, three energy management strategies included a fuzzy logic control (FLC), an equivalent consumption minimization strategy (ECMS) and a state machine strategy based on droop control (SMS-DC) are utilized to coordinate and distribute the power demand from each power source of the tramway, and the comparisons are also carried out to verify the validity according to a driving cycle.

2. Configuration of hybrid tramway

A hybrid tramway without grid connection called LF-LRV is being developed by Chinese manufacturer of Tangshan Railway Vehicle Co. Ltd and Southwest Jiaotong University. This hybrid tramway, which presents a capacity of 360 passengers and reaches a maximum speed of 77 km/h, is composed of two motor units and one trailer unit. Two motor units are supplied by the tramway traction system via an inverter box. The proposed configuration of PEMFC–battery–SC powered hybrid tramway is shown in **Figure 1**. The hybrid powertrain is composed of the PEMFCs (manufactured by Ballard), the batteries (manufactured by Microvast Power), the SCs (manufactured by Maxwell), the unidirectional DC/DC converters, the bidirectional DC/DC converters, the traction motors, auxiliary service module, and braking resistor. The PEMFC systems (No. 1 and No. 2) and two unidirectional DC/DC converters are placed on the trailer unit. The batteries and the SCs (No. 1 and No. 2) and two bidirectional DC/DC converters are, respectively, placed on two motor units. The boost-type unidirectional DC/DC converters are used to raise the DC voltage to 750 V. Two batteries and two SCs are connected to the traction DC bus through two bidirectional DC/DC converters, respectively, which allow the charge and discharge of the batteries and the SCs. Particularly, two SCs are utilized to consume the peak power that neither the PEMFCs nor the batteries can store because of their high specific power and high dynamic response.

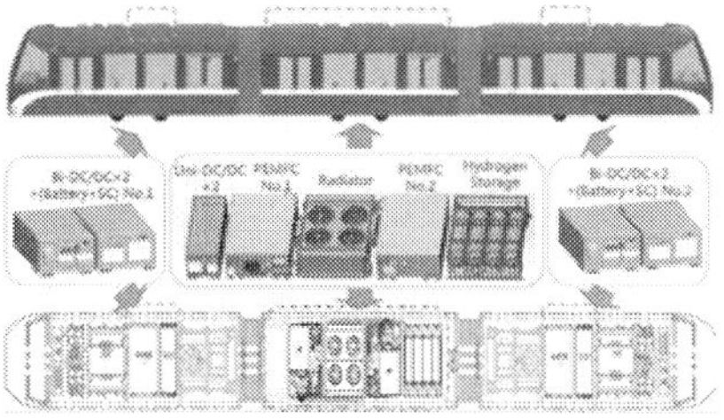

Figure 1. Configuration of PEMFC–battery–SC powered hybrid tramway.

3. Modeling of hybrid power system

3.1. Modeling of PEMFC power unit

A PEMFC power unit, which consists of a power module subsystem, a control subsystem, and a hydrogen storage subsystem, is considered as the heart of the hybrid power system for the tramway [6, 8]. The PEMFC power unit is setup in the clean energy laboratory of Southwest Jiaotong University. In the designed PEMFC power unit, a Ballard Stack HD6 Module-FCvelocity™ is used as the PEMFC stack module. This stack module is rated at 150 kW gross power, which contains the auxiliary components for hydrogen recirculation and purge, and air humidification. The key component for the air delivery module is the air compressor, which utilizes a turbo charger manufactured by ROTREX™ Corporation. The cooling module is fulfilled with two separate cooling loops. The primary cooling loop provides heat rejection for the HD6 Module. The second cooling subsystem provides heat rejection for the PEMFC stack module condenser to ensure enough process water available at all times for air humidification.

A model of PEMFC power unit is developed based on the 150 kW Ballard HD6 Module testing data. A set of assumptions generally undertaken in similar studies is carried out to focus on the dynamic response analysis for the HD6 Module requested current and output voltage [15–19]. The PEMFC output voltage is decreased from its equilibrium potential because of irreversible losses. Particularly, the concentration losses are neglected under practical working condition. The output voltage equation is expressed as

$$U_{fc} = E_{oc} - U_{act} - U_{ohmic} \tag{1}$$

with

$$\begin{cases} E_{oc} = K_c\left[E_0 + (T-298)\dfrac{-44.43}{zF} + \dfrac{RT}{zF}\ln(P_{H_2}P_{O_2}^{1/2})\right] \\ U_{act} = \dfrac{1}{\tau s+1}NA_{nom}ln(\dfrac{i_{fc}}{i_0}) \\ U_{ohmic} = R_{internal}i_{fc} \end{cases} \tag{2}$$

where E_{oc} is the open circuit voltage, U_{fc} is the stack output voltage, U_{ohmic} is the ohmic voltage drop, and U_{act} is the activation voltage drop, E_0 is the electromotive forces under standard pressure, K_c rated voltage constant, T is the operating temperature, z is transfer electron number, P_{H2} and P_{O2} are the gas pressures, F is Faraday constant, R is the gas constant, i_{fc} is the cell output current, $R_{internal}$ is inner resistance of a stack, N is the number of cells, τ is the dynamic response time constant that will make the cell output voltage exhibits a delay approximately during a sudden change in cell output current. The detailed description about the modeling of the PEMFC power unit based on HD6 Module can be found in [6, 8]. In order to prove the validity of the PEMFC power unit model, the comparisons between the experimental data and the characteristics curve of polarization by simulation are achieved. The polarization curve by simulation and the experimental data are in good match as shown in **Figure 2**.

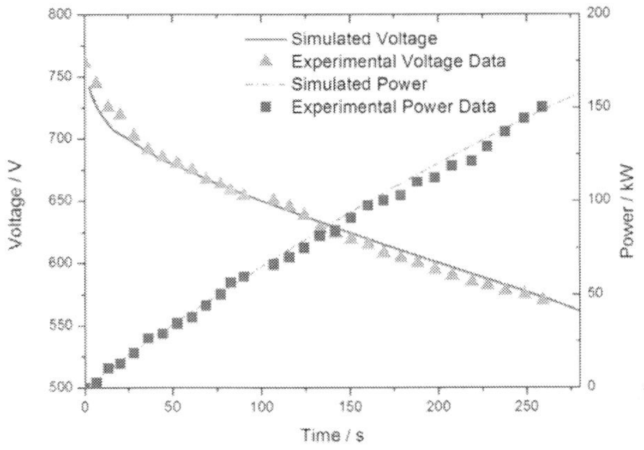

Figure 2. Comparisons between the experimental data and the polarization curve by simulation.

3.2. Modeling of energy store system

The batteries considered for the hybrid system are of type Li-ion as they have proven to exhibit high energy density and efficiency and a large number of charge/discharge cycles compared with other battery types [6, 7, 10]. The Li-ion batteries are utilized both for supplying a portion of the base load together with PEMFC and capturing the braking energy together with the SC. The behavior of this battery is represented by a modified Shepherd curve-fitting model presented in [10, 20], which is available in the SimPowerSystems toolbox of Simulink. In this model, a voltage polarization is added to the battery discharge voltage expression to better represent the effect of the battery stage of charge (SOC) on the battery performance. The validity of this model has been verified by experimental studies [20–22]. The battery SOC is calculated as follows:

$$SOC = 100[1 - \frac{1}{Q}\int_0^t i_{\text{bat}}(t)dt] \tag{3}$$

According to the battery open circuit voltage and voltage drop resulting from the battery-equivalent internal impedance, the battery output voltage can be calculated as follows:

$$U_{\text{bat}} = E_0 - K\frac{Q}{Q-it}it - R_{\text{b}}i_{bat} + A_{\text{b}}\exp(-Bit) - Pol_{\text{res}}i^* \tag{4}$$

where E_0 is the constant voltage, i_{bat} is the battery current, K is the polarization constant, Q is the maximum battery capacity, i^* is the filtered battery current, it is the actual battery charge, A_b is the exponential zone amplitude, B is the exponential zone time constant inverse, and R_b is the battery internal resistance. Pol_{res} is the polarization resistance, which can be decided as follows:

$$Pol_{res} = K\frac{Q}{Q-it}(1-u(t)) + K\frac{Q}{it-0.1Q}u(t) \tag{5}$$

where u(t) is equal to 1 if battery is charging and is equal to 0 if battery is discharging. The detailed description about the modeling of the battery model can be found in [6, 7, 20].

The SC, which is known as electric double-layer capacitor, is similar to conventional electrostatic or electrolytic capacitor with the advantage that it can store or release more energy due to their high capacitance. The SC model presented in [20, 23] is available in the SimPowerSystems toolbox of Simulink. This model consists of a capacitance representing the SC performance during the charging and discharging process and an equivalent series resistance representing the charging and discharging resistance. The SC output voltage is expressed considering resistive losses as follows:

$$U_{sc} = \frac{Q_T}{C_T} - R_{sc}i_{sc} \tag{6}$$

$$\text{with} \quad \begin{cases} C_T = \dfrac{N_p}{N_s}C \\ Q_T = N_pQ_c = \int i_{sc}dt \\ C = (\dfrac{1}{C_H} + \dfrac{1}{C_{GC}})^{-1} \\ C_H = \dfrac{N_e\varepsilon\varepsilon_0 A_i}{d} \\ C_{GC} = \dfrac{FQ_c}{2N_eRT}\sinh(\dfrac{Q_c}{N_e^2A_i\sqrt{8RT\varepsilon\varepsilon_0 c}}) \end{cases} \tag{7}$$

where C_T is the total capacitance, Q_T is the total electric charge, R_{SC} is the SC internal resistance, i_{SC} is the SC current, N_s is the number in series, N_p is the number in parallel, C_H and C_{GC} are the Helmholtz and Gouy–Chapman capacitance, N_e is the number of electrode layers, ε and ε_0 are the permittivity values of the electrolyte material and free space, A_i is the interfacial area between electrodes and electrolyte, d is the Helmholtz layer length, Q_c is the cell electric charge, and c is the molar concentration. In order to validate the SC model, a comparison test with commercial Maxwell BMOD SC, which is specifically designed for high-power transport applications such as locomotive and the tramway, has been carried out in [20]. The validity of this model has been verified by experimental studies. The detailed description about the modeling of the SC model can be found in [6, 7, 20].

3.3. Modeling of DC/DC converters

A power electronic device is composed of a PWM-based DC/DC converter for each power source, which connects the sources with the DC bus. The PEMFC, battery, and SC present a lower terminal voltage than the DC voltage necessary to feed the traction inverter. Two unidirectional boost DC/DC converters have been developed by using the two-quadrant chopper models included in SimPowerSystems of Simulink [24, 25]. These converter connects two PEMFC systems with the 750VDC bus maintaining the PEMFC systems stable despite variations in load. Two bidirectional converters are, respectively, utilized to the batteries and the SCs with boost operation if discharging and buck operation if charging. These converter models also been developed by using the two-quadrant chopper models included in SimPowerSystems [24, 25]. The batteries and the SCs are located on the low-voltage side, and the high-voltage side is connected to the 750VDC bus. Therefore, the effective control of the duty cycle of the bidirectional converters assure the rated voltage of the DC bus and the charge and discharge safety of the batteries and the SCs.

4. Energy management strategies for hybrid tramway

4.1. A state machine strategy based on droop control

A droop control approach is presented to coordinate multiple motor and trailer units (multiple power sources), which is responsible for determining the reference power command to match the load requests from a driving cycle of tramway [7]. A reference power P_{ref} versus bus voltage U_{bus} droop is adopted. This characteristic is designed to convert the external command voltage U_{dcref} = 750 V into the reference power P_{ref}. The larger is the amount of reference power that is injected or absorbed, the lower of the bus voltage is allowed to sag compared to the external request. Conversely, P_{ref} is allowed to swell as the power is injected or absorbed. This correction successfully limits the circulating currents because it limits the power injections or absorptions to achieve the adjusted voltages. It is responsible for modifying the value of the bus voltage. The droop characteristic is realized by multiplying the droop coefficient m with the voltage error for the reference power, which is expressed as follows:

$$P_{ref} = \begin{cases} P_{reflimt1} & U_{bus} > U_{H2} \\ m(U_{bus} - U_{dcref} - \Delta U) & U_{H1} < U_{bus} < U_{H2} \\ 0 & U_{L1} < U_{bus} < U_{H1} \\ m(U_{dcref} - \Delta U - U_{bus}) & U_{L2} < U_{bus} < U_{L1} \\ P_{reflimt2} & U_{bus} < U_{L2} \end{cases} \tag{8}$$

where U_{L2} and U_{H2} are upper and lower voltage threshold based on $P_{reflimit1}$ and $P_{reflimit2}$, and U_{L1} and U_{H1} are upper and lower voltage threshold of active power sources. The droop coefficient m determines the slopes of the droop characteristics, and it depends on the values of m_a, m_b, k_1, k_2, SOC_{bat_i}, and SOC_{sc_i} ($i = 1\ldots n$, n is the numbers of batteries or SC), which is defined as follows:

$$m = \begin{cases} \dfrac{m_a}{k_1 \dfrac{\sum_{i=1}^{n} SOC_{bat\_i}}{n} + k_2 \dfrac{\sum_{i=1}^{n} SOC_{sc\_i}}{n}} & U_{H1} < U_{bus} < U_{H2} \\ m_b \left(k_1 \dfrac{\sum_{i=1}^{n} SOC_{bat\_i}}{n} + k_2 \dfrac{\sum_{i=1}^{n} SOC_{sc\_i}}{n}\right) & U_{L1} < U_{bus} < U_{L2} \end{cases} \tag{9}$$

Furthermore, the state machine strategy based on droop control (SMS-DC) is developed for the hybrid tramway. The aim of the SMS-DC is to decide the PEMFC reference power with the state change. Five states designed by the SMS are defined to distribute the reference power signals P_{fcref}, P_{batref}, and P_{scref} for the PEMFC, the battery and the SC through regulating two unidirectional DC/DC converters, and four bidirectional DC/DC converters. SOC_{batmin} and SOC_{batmax} are the lower and the upper limits of battery SOC; SOC_{scmin} and SOC_{scmax} are the lower and the upper limits of SC SOC; $SOC_{batnom1}$ and $SOC_{batnom2}$ are the upper and lower bounds of the preferred zone of battery SOC; and SOC_{scnom1} and SOC_{scnom2} are the upper and lower bounds of the preferred zone of SC SOC. And also, the SOC levels of batteries and SCs are divided into the high SOC, normal SOC and low SOC, and the changes between these levels are performed by means of two hysteresis cycles. The state chart diagram of the SMS is described as **Figure 3** shown. The PEMFC reference power P_{fcref} is determined based on the batteries and SCs SOC range and the reference power derived from droop control. In addition, the reference signal is decided for energy dissipation via the braking resistor if required during regenerative braking. These states are defined as follows:

State 1:

$$P_{fcref\_i} = \begin{cases} P_{fcmax} & if\ P_{ref} \geq nP_{fcmax} + P_{aux} \\ P_{ref}/n & if\ nP_{fcmax} + P_{aux} > P_{ref} \geq nP_{fcmin} + P_{aux} \\ P_{fcmin} & if\ nP_{fcmin} + P_{aux} > P_{ref} \end{cases} \tag{10}$$

State 2:

$$P_{fcref\_i} = P_{fcpre\_i} \tag{11}$$

State 3:

$$P_{\text{fcref\_i}} = \begin{cases} P_{\text{fcmax}} & \text{if } P_{\text{ref}} \geq nP_{\text{fcmax}} + P_{\text{aux}} \\ P_{\text{ref}}/n & \text{if } nP_{\text{fcmax}} + P_{\text{aux}} > P_{\text{ref}} \geq nP_{\text{fcopt}} + P_{\text{aux}} \\ P_{\text{fcopt}} & \text{if } nP_{\text{fcopt}} + P_{\text{aux}} > P_{\text{ref}} \end{cases} \tag{12}$$

State 4:

$$P_{\text{fcref\_i}} = P_{\text{fcpre\_i}} \tag{13}$$

State 5:

$$P_{\text{fcref\_i}} = \begin{cases} P_{\text{fcmax}} & \text{if } P_{\text{ref}} \geq nP_{\text{fcmax}} + P_{\text{aux}} \\ P_{\text{ref}}/n + P_{\text{char}} & \text{if } nP_{\text{fcmax}} + P_{\text{aux}} \geq P_{\text{ref}} \end{cases} \tag{14}$$

where $P_{\text{fcref\_i}}$ denotes the reference power of *i*-th PEMFC system, P_{fcmin} and P_{fcmax} are the lower and upper limits of PEMFC system, P_{aux} is the auxiliary services power of tramway, P_{char} is the minimum charging power of battery and SC, *and* P_{fcopt} is the optimal power of PEMFC system.

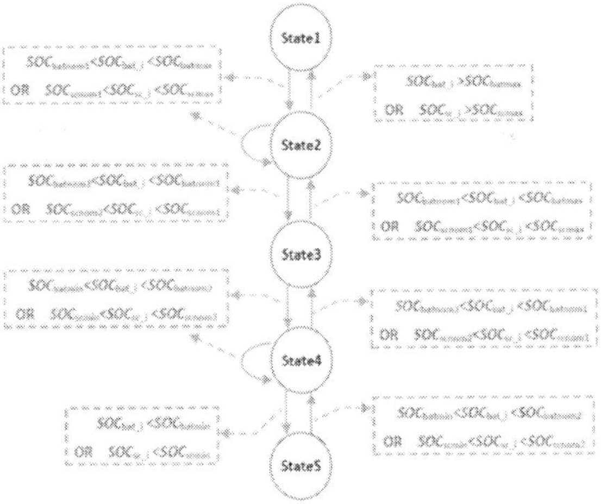

Figure 3. State chart diagram of the SMC-DC.

4.2. Energy management strategy based on fuzzy logic control

The fuzzy logic controller (FLC) is more appropriate for the energy management system of hybrid tramway. A list of IF-THEN rules is adopted to describe input and output relationship of the controller [6]. In the fuzzy inference system, two decentralized FLC are designed improve the efficiencies of tramway and PEMFC systems on condition that dynamic property of tramway is satisfied.

The No. 1 FLC (FLC1) has three input variables and one output variable. The input variables include the demand power of electrical motor P_{m1}, the battery SOC, and the SC bank SOC (CSOC). The PEMFC reference power P_{ref1} is defined as the output variable. The No. 2 FLC (FLC2) has three input variables and one output variable. The surplus demand power of motor P_{m2}, and the battery SOC and the SC CSOC are defined as the input variables. The SC reference power P_{ref3} is defined as the output variable. The form of fuzzy reasoning is "IF P_{m1} is A_i AND SOC is B_i AND CSOC is C_i, Then P_{ref1} is D_i". Mamdani inference method is adopted to implement the fuzzy reasoning. More detailed description about the energy management strategy based on fuzzy logic control can be found in [6].

4.3. Equivalent consumption minimization strategy

In the present hybrid tramway, if the batteries or the SCs supply electrical energy, their SOC decreases, so that the batteries or the SCs will need to be recharged by the energy proceeded from two PEMFC systems or the tramway braking in order to maintain a desired SOC. Hence, extra hydrogen consumption could be needed due to the energy obtained from the tramway braking is generally insufficient. If the batteries or the SCs is recharged their SOC increases and the energy will supply to the tramway in future accelerations or start-ups resulting in a reduction of the hydrogen consumption. The electrical energy consumption of the batteries and the SCs are transformed into an equivalent hydrogen consumption to make the two comparable [10].

The equivalent consumption minimization strategy (ECMS) focuses on minimizing the equivalent hydrogen consumption of the hybrid powertrain C (g), which is calculated as sum of two PEMFC systems hydrogen consumption C_{fc}, and two batteries and two SCs-equivalent hydrogen consumptions, C_{bat} and C_{sc}. In this work, the mathematical problem used to optimize the equivalent hydrogen consumption is formulated as follows:

$$min\ C = min\ \left(C_{fc} + \omega_1 C_{bat} + \omega_2 C_{sc}\right)$$

$$subject\ to \begin{cases} SOC_{batmin} \le SOC_{bat\_i} \le SOC_{bat\,max} \\ SOC_{sc\,min} \le SOC_{sc\_i} \le SOC_{sc\,max} \\ P_{reflimt1} \le P_{ref} \le P_{reflmit2} \end{cases} \tag{15}$$

where ω_1 and ω_2 is the penalty coefficients, which modifies the batteries and the SCs-equivalent hydrogen consumption up or down depending on their SOC deviation from the target. This optimization problem is subject to the constrains of the batteries and the SCs charging-sustaining and the power level. In addition, since the aim of two SCs is to provide the power peaks demanded by the hybrid tramway during the acceleration or braking that two PEMFC systems and two batteries cannot generate or absorb, their average value will be minimum and C_{sc} can be neglected compared with C_{fc} and C_{bat}. Hence, ω_2 is set to zero here, and ω_1 limited by μ and the boundary of SOC is expressed as follows:

$$\omega_1 = 1 - 2\mu \frac{\left[SOC_{bat\_i} - 0.5\left(SOC_{batmax} + SOC_{batmin} \right) \right]}{SOC_{batmax} - SOC_{batmin}} \tag{16}$$

The battery-equivalent hydrogen consumption C_{bat} can be calculated from the battery power P_{bat}. This concept is developed in [10, 11], where the battery charging and discharging efficiencies are calculated based on the equivalent internal resistance model. The battery-equivalent hydrogen consumption C_{bat} can be expressed as follows:

$$C_{bat} = \begin{cases} \dfrac{P_{bat}}{\eta_{chavg}} \dfrac{C_{fcavg}}{P_{fcavg}} \left(\dfrac{1}{2} + \dfrac{1}{2}\sqrt{1 - \dfrac{4R_{dis}P_{bat}}{E^2}} \right)^{-1} & P_{bat} \geq 0 \\ P_{bat}\eta_{disavg} \dfrac{C_{fcavg}}{P_{fcavg}} \left(\dfrac{1}{2} + \dfrac{1}{2}\sqrt{1 - \dfrac{4R_{ch}P_{bat}}{E^2}} \right)^{-1} & P_{bat} < 0 \end{cases} \tag{17}$$

where η_{chavg} and η_{disavg} are the mean efficiencies of the battery charging and discharging, respectively, C_{fcavg} is the average hydrogen consumption of the PEMFC systems, P_{fcavg} is the average power of the PEMFC systems, *and* R_{ch} and R_{dis} are the resistances of the battery charging and discharging, respectively. The analytic solution to the optimized problem defined in Eq. (16) can be expressed as follows:

$$P_{batop} = \begin{cases} U_{batmin}\left(E - U_{batmin}\right)/R_{dis} & K_3 \leq x_{min} \\ E^2\left(1 - K_3^2\right)/4R_{dis} & x_{min} < K_3 \leq 1 \\ 0 & 1 < K_3 \leq 1/\left(\eta_{chavg}\eta_{disavg}\right) \\ E^2\left[1 - \left(K_3\eta_{chavg}\eta_{disavg}\right)^2\right]/4R_{dis} & 1/\left(\eta_{chavg}\eta_{disavg}\right) < K_3 \leq x_{max}/\left(\eta_{chavg}\eta_{disavg}\right) \\ -U_{batmax}\left(U_{batmax} - E\right)/R_{ch} & K_3 \geq x_{max}/\left(\eta_{chavg}\eta_{disavg}\right) \end{cases} \tag{18}$$

where P_{batopt} is the optimized power of the battery, *and* U_{batmin} and U_{batmax} are the minimal and the maximal allowed battery voltages, respectively. K_1, x_{min}, *and* x_{max} are defined variables. In addition, since the aim of two SCs is to provide the power peaks demanded by the hybrid tramway during the acceleration or braking that two PEMFC systems and two batteries cannot generate or absorb, their average value will be minimum and C_{sc} can be neglected compared with C_{fc} and C_{bat}.

5. Results and discussions

A real driving cycle of the tramway from Samsun in Turkey is adopted to evaluate the performance of these energy management strategies for the hybrid system. A round-trip route, which consists of a symmetrical route of going and return, has been adopted as shown in **Figure 4**. The hybrid propulsion system of tramway is composed of commercial devices, such as Ballard FCvelocity™ HD6, Microvast Battery MV06203127NTPCA, and Maxwell BMOD0615. The premises are considered carefully for the hybrid system sizing. The maximum power of two PEMFC systems should be higher than the average power demanded by the tramway during the driving cycle in order to avoid excessive SOC drop of the batteries and the SCs. A hybrid propulsion system considered for the tramway presents two PEMFC systems (300 kW), two Li-ion batteries (40 Ah), and two SC banks (45 F). These devices have been selected from commercially available components as **Table 1** shown.

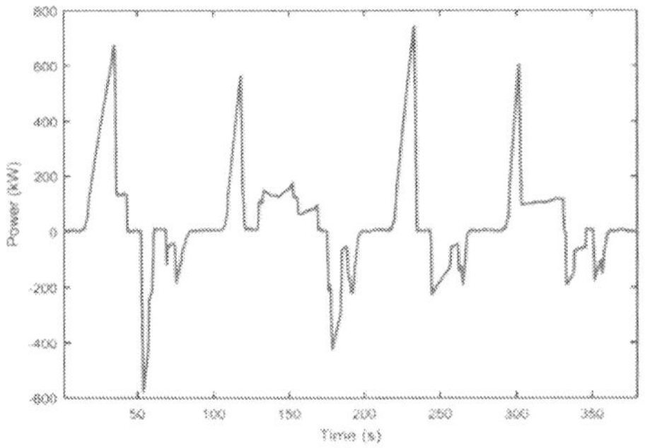

Figure 4. Driving cycle of hybrid tramway.

PEMFC			
Manufacturer	Ballard stack modules-FCvelocity™ HD6		
Rated power (kW)	260	Rated voltage range (V)	570–650
Nominal coolant temperature (°C)	50-63	Mass (kg)	710
Maximum coolant temperature (°C)	66	Set numbers	2 parallel
Li-ion Battery			
Manufacturer	Microvast battery MV06203127NTPCA		
Capacity (Ah)	40	Rated voltage (V)	331
Maximum discharging rate (C)	5	Set numbers	2 parallel
Maximum discharging current (a)	240	Mass (kg)	280
Supercapacitor (SC)			
Manufacturer	Maxwell BMOD0615		
Capacity (F)	45	Rated voltage (V)	528
Absolute maximum current (A)	1400	Specific power (W/kg)	3300
Set numbers	(11 series * 3 parallel) * 2 parallel	Mass (kg)	670

Table 1. Parameters of commercial devices.

The performance comparison of these energy management strategies are carried out under the same the driving cycle and hybrid topology. In **Table 2**, the criteria for performance comparison are the average efficiency and equivalent hydrogen consumptions of tramway, the average value of SOC_{bat} and SOC_{sc}, and the variation range of SOC_{bat} and SOC_{sc}. Compared with the other strategies, the SMS-DC provides better tramway average efficiency (56.78%) and the tramway-equivalent hydrogen consumptions (301.22 g). The maximum difference obtained between the best and the worst tramway average efficiency is 7.7%. The efficient use of the batteries and SCs energy (average SOC of 75.12 and 78.02%) are also achieved by the SMS-DC because the final average SOC value of batteries and SCs are closer to the initial value. The best results related to the variation ranges of SOC_{bat} and SOC_{sc} are obtained by the ECMS, but at the expense of lower overall efficiency (49.72%).

EMS			
Criteria	**FLC**	**ECMS**	**SMS-DC**
Tramway average efficiency (%)	49.08	49.72	56.78
Tramway-equivalent hydrogen consumptions (g)	413.69	325.71	301.22
Average SOC_{bat} (%)	73.37	73.56	75.12
Average SOC_{sc} (%)	79.08	77.92	78.02
SOC_{bat} range	[70, 77]	[73, 76]	[72, 76]
SOC_{sc} range	[55, 92]	[63, 80]	[57, 82]

Table 2. Results of different strategies.

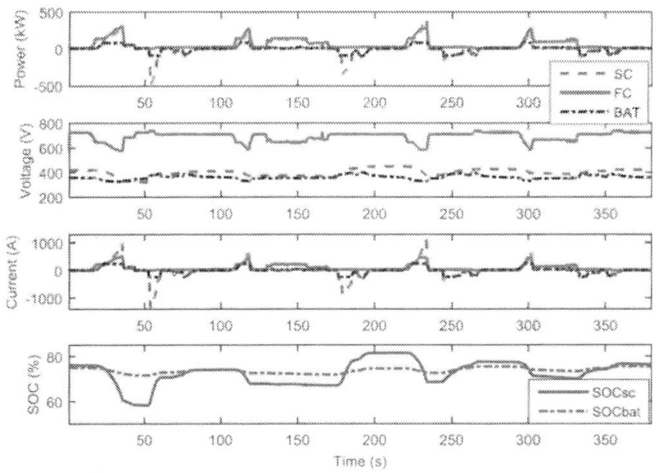

Figure 5. Output power, voltage, current, and SOC of different power sources.

As the results of the energy management system based on the SMS-DC, the output power, voltage, current, and SOC of different power sources are given in **Figure 5**, respectively. With regard to the output power, the PEMFC systems only increase or decrease the power during high accelerations or brakings. It can be observed that the proposed SMS based on droop control is able to guarantee the stable operation of the PEMFC systems during the most of the drive cycle. The output power of the batteries alternates between negative and positive according to charging or discharging. It helps provide a portion of the positive low-frequency components of power demand to reduce the burden of the PEMFC and absorb the slow-variation negative portion. Additionally, due to the PEMFC dynamic limitation, the SCs provide the transient power demand successfully during sudden acceleration or braking. Their fast response fulfill the power demand, increase the hybrid system power density, and have to generate or absorb the power, which, either the PEMFC or the battery, are not able to generate or absorb.

6. Conclusions

In this chapter, the energy management system for the hybrid tramway based on multiple PEMFC systems, batteries, and SCs is designed without grid connection. The hybrid propulsion system model is developed with commercially available devices. The FLC, the ECMS, and the SMS-DC are utilized to coordinate and distribute the power demand to each power source appropriately, avoid the transients and rapid changes of power demand, and achieve high efficiency. According to the driving cycle of the tramway, the testing of the energy management systems is carried out. The comparisons with the FLC, the ECMS, and the SMS-DC verify that the higher average efficiencies of the tramway, the lower tramway-equivalent hydrogen consumptions, and more efficient use of the batteries and SCs energy are achieved by the SMS-DC. Hence, the suitable designed energy management system for the hybrid tramway will enhance the hydrogen consumptions of overall hybrid system and the efficiencies of each power source.

Acknowledgements

This work was supported by National Natural Science Foundation of China (51177138, 61473238, 51407146), National Key Technology R&D Program (2014BAG08B01), and Sichuan Provincial Youth Science and Technology Fund (2015JQ0016).

Author details

Qi Li* and Weirong Chen

*Address all correspondence to: liqi0800@163.com

School of Electrical Engineering, Southwest Jiaotong University, Chengdu, China

References

[1] O. Erdinc, M. Uzunoglu. Recent trends in PEM fuel cell-powered hybrid systems: investigation of application areas, design architectures and energy management approaches. Renew. Sustain. Energy Rev. 2010;14:2874–2884. doi:10.1016/j.rser.2010.07.060.

[2] C. Bao, M. Ouyang, B. Yi. Modeling and control of air stream and hydrogen flow with recirculation in a PEM fuel cell system-II. Linear and adaptive nonlinear control. Int. J. Hydrogen Energy. 2006;31:1897–1913. doi:10.1016/j.ijhydene.2006.02.030.

[3] Q. Li, W. Chen, Y. Li, S. Liu, J. Huang. Energy management strategy for fuel cell/battery/ultracapacitor hybrid vehicle based on fuzzy logic. J. Electrical Power Energy Syst. 2012;43:514–525. doi:10.1016/j.ijepes.2012.06.026.

[4] J. Jia, G. Wang, Y.T. Cham, Y. Wang, M. Han. Electrical characteristic study of a hybrid PEMFC and ultracapacitor system. IEEE Trans. Ind. Electron. 2010;57:1945–1953. doi:10.1109/TIE.2009.2022066.

[5] L. Xu, J. Li, M. Ouyang, J. Hua, G. Yang. Multi-mode control strategy for fuel cell electric vehicles regarding fuel economy and durability. Int. J. Hydrogen Energy. 2014;39:2374–2389. doi:10.1016/j.ijhydene.2013.11.133.

[6] Q. Li, W. Chen, Z. Liu, M. Li, L. Ma. Development of energy management system based on a power sharing strategy for a fuel cell–battery–supercapacitor hybrid tramway. J. Power Sources. 2015;279:267–280. doi:10.1016/j.jpowsour.2014.12.042.

[7] Qi Li, Hanqing Yang, Ying Han, Ming Li, Weirong Chen. A State Machine Strategy based on Droop Control for an Energy Management System of PEMFC-Battery-Supercapacitor Hybrid Tramway. Int. J. Hydrogen Energy. 2016;41: 12705–12713. DOI: 10.1016/j.ijhydene.2016.04.254.

[8] F. Peng, W. Chen, Z. Liu, Q. Li, C. Dai. System integration of China's first proton exchange membrane fuel cell locomotive. Int. J. Hydrogen Energy. 2014;39:13886–13893. doi:10.1016/j.ijhydene.2014.01.166.

[9] A.R. Miller, K.S. Hess, D.L. Barnes, T.L. Erickson. System design of a large fuel cell hybrid locomotive. J. Power Sources. 2007;173:935–942. doi:10.1016/j.jpowsour.2007.08.045.

[10] P. García, J.P. Torreglosa, L.M. Fernández, F. Jurado. Viability study of a FC–battery–SC tramway controlled by equivalent consumption minimization strategy. Int. J. Hydrogen Energy. 2012;37:9368–9382. doi:10.1016/j.ijhydene.2012.02.184.

[11] J.P. Torreglosa, F. Jurado, P. Garcı´a, L.M. Ferna´ndez. Hybrid fuel cell and battery tramway control based on an equivalent consumption minimization strategy. Control Eng. Pract. 2011;19:1182–1194. doi:10.1016/j.conengprac.2011.06.008.

[12] L.M. Fernandez, P. Garcia, C.A. Garcia, F. Jurado. Hybrid electric system based on fuel cell and battery and integrating a single dc/dc converter for a tramway. Energy Convers. Manag. 2011;52:2183–2192. doi:10.1016/j.enconman.2010.12.028.

[13] P. Thounthong, P. Tricoli, B. Davat. Performance investigation of linear and nonlinear controls for a fuel cell/supercapacitor hybrid power plant. J. Electrical Power Energy Systems. 2014;54:454–464. doi:10.1016/j.ijepes.2013.07.033.

[14] Y. Eren, O. Erdinc, H. Gorgun, M. Uzunoglu, B. Vural. A fuzzy logic based supervisory controller for an FC/UC hybrid vehicular power system. Int. J. Hydrogen Energy. 2009;34:8681–8694. doi:10.1016/j.ijhydene.2009.08.033.

[15] Q. Li, W. Chen, Z. Liu, A. Guo, J. Huang. Nonlinear multivariable modeling of locomotive proton exchange membrane fuel cell system. Int. J. Hydrogen Energy. 2014;39:13777–13786. doi:10.1016/j.ijhydene.2013.12.211.

[16] Q. Li, W. Chen, Z. Liu, A. Guo, S. Liu. Control of proton exchange membrane fuel cell system breathing based on maximum net power control strategy. J. Power Sources. 2013;214:212–218. doi:10.1016/j.jpowsour.2013.04.067.

[17] J.T. Pukrushpan, A.G. Stefanopoulou, H. Peng. Control of fuel cell breathing. IEEE Control Syst. Mag. 2004;24:30–46. doi:10.1109/MCS.2004.1275430.

[18] A. Arce, A.J. del Real, C. Bordons, D.R. Ramirez. Real-time implementation of a constrained MPC for efficient airflow control in a PEM fuel cell. IEEE Trans. Ind. Electron. 2010;57:1892–1905. doi:10.1109/TIE.2009.2029524.

[19] Q. Li, W. Chen, Y. Wang, J. Jia, M. Han. Nonlinear robust control of proton exchange membrane fuel cell by state feedback exact linearization. J. Power Sources. 2009;194:338–348. doi:10.1016/j.jpowsour.2009.04.077.

[20] S.N. Motapon, L.A. Dessaint, K. Al-Haddad. A comparative study of energy management schemes for a fuel-cell hybrid emergency power system of more-electric aircraft. IEEE Trans. Ind. Electron. 2014;61:1320–1334. doi:10.1109/TIE.2013.2257152.

[21] T.C. Yang. Initial study of using rechargeable batteries in wind power generation with variable speed induction generators. IET Renew. Power Gener. 2008;2:89–101. doi: 10.1049/iet-rpg:20070008.

[22] G. Sikha, R.E. White, B.N. Popov. A mathematical model for a lithium-ion battery/ electrochemical capacitor hybrid system. J. Electrochem. Soc. 2005;152:1682–1693. doi: 10.1149/1.1940749.

[23] K.B. Oldham. A Gouy-Chapman-Stern model of the double layer at a (metal)/(ionic liquid) interface. J. Electroanal. Chem. 2008;613:131–138. doi:10.1016/j.jelechem. 2007.10.017.

[24] M. Delshad, H. Farzanehfard. A new soft switched push pull current fed converter for fuel cell applications. Energy Convers. Manag. 2011;52:232–243. doi:10.1016/j.enconman.2010.08.019.

[25] D.D. Marquezini, D.B. Ramos, R.Q. Machado, F.A. Farret. Interaction between proton exchange membrane fuel cells and power converters for AC integration. IET Renew. Power Gener. 2008;2:151–161. doi:10.1049/iet-rpg:20070057.

Chapter 3

Wireless Communication Systems for Urban Transport

Juan Moreno García-Loygorri,

José Manuel Riera and Carlos Rodríguez

Additional information is available at the end of the chapter

http://dx.doi.org/10.5772/65585

Abstract

This chapter describes the main features of the wireless communication systems of urban rail and related applications. The perspective will be complete: application, network and physical layers will be discussed. Moreover, to properly address some of the challenges that these systems face, we will provide a deep insight into propagation issues related to tunnels and urban areas. Finally, a detailed survey on the directions of research on all these topics will be provided.

Keywords: communications, LTE, mass transit, transport, urban, wireless, WLAN

1. Introduction

In the last few years, wireless communication systems for transport systems, in particular for urban transport, are gaining importance. The so-called intelligent transportation systems have a major impact on the performance of transport, because they can increase the capacity of the lines (and even add some extra safety), provide added-value features to operators (i.e., remote maintenance) and also additional services to passengers, such as the very popular access to the Internet. However, commissioning all these features implies facing many relevant challenges at every conceivable layer, not only technical, but also regulatory, financial, etc. In this chapter, we have adopted a practical approach to all these issues, trying to focus on what the railway engineer really needs to know: the essentials, but also the challenges and where the Gordian knot is.

In this chapter, starting from the application layer and ending on the physical one, we cover the most important aspects of these systems. The chapter structure is based on the well-known open systems interconnection (OSI) stack and, in our opinion, is very helpful to discuss all the details of very complex systems such as wireless in urban transport. We start in Section 2 by presenting the most relevant systems from the point of view of the criticality of the service: safety services such as communications-based train control (CBTC) or public safety (PS) radios; operator-oriented, like closed-circuit television (CCTV) or operation and maintenance (O&M) systems; and, finally, services only focused on the passenger (with no safety consequences), like infotainment or the Internet access. In Section 3, we provide a deep insight on networks and technologies that make all the previous services possible. In this section we discuss some aspects that require our attention when deploying services such as these, with all the mobility issues and many other problems that may arise. Section 4 is dedicated to explain the physical layer related to urban transport. Two main scenarios need to be covered: strict-sense urban (city landscape, mostly for tramways) and tunnels (mostly for subways). These two scenarios present significant differences between them, tunnels especially implying a challenge for wireless systems. In this section we also discuss the 'spectrum problem' in order to provide some key ideas on the frequency allocation of wireless systems. 'Wireless communication systems for urban transport' is a fast-evolving field and, as mentioned before, it is far from being finished; and a lot of research work is still in progress. The European Union, through its H2020 initiatives (particularly, Shift2Rail [1] and Roll2Rail [2]) is investing a lot of funding in some railway 'hot topics'. In Section 5 we introduce all the lines of research that are meant to change the face of railways in the next 20 years. Finally, Section 6 hosts the main conclusions of this chapter.

2. Urban rail applications based on wireless communications

The next step in this chapter is to know about the railway services that demand wireless connectivity. This is the most visible part for railway operators, stakeholders and any other railway-related agent. Service must be understood as the supply of something needed. Therefore it is a synonym of application and in the current text we will use both of them interchangeably.

There is no general consensus on the classification of services, but railways always have paid a lot of attention to safety, so we use this approach to classify services. Accordingly we may say that we have three types of services: mission-critical (safety-related); operator-oriented, with no safety implications but operators heavily rely on them (like CCTV); and finally customer-oriented: every service that is intended only for the passengers, having no relation with operational or safety issues. Sometimes it is very complex to state clearly which domain a function belongs to. For example, consider a CCTV camera placed in the front of a train which is used by an operator to detect if somebody has fallen onto the track (letting the driver to stop the train before running over the individual). In strict terms, CCTV is an operator service (with no safety implications) but if this system fails, it could cost lives.

http://dx.doi.org/10.5772/65585

2.1. Mission critical services

Two taxonomies fell within this category: signaling-related, with a strong safety commitment and public-safety systems. The main difference between the two of them is that signaling systems take care of the safe movement of the train (telling the train how far it can go and how fast; helping the train to report its position and speed to the wayside equipment; etc.) and public-safety communication systems that do not control any movement of the train but provide a safe communication channel that enables railway agents (drivers, operational control center (OCC) controllers, maintenance and security staff, etc.) to communicate appropriately.

Railway signaling is a very broad field whose extent exceeds by far the purpose of this chapter. Here we focus on the implications on wireless communication systems that carry the data of the signaling service. For urban rail, the most relevant transmission-based signaling system is CBTC (communication-based train control). Using the definition of the very CBTC standard [3], this system is "a continuous automatic train control system utilizing high-resolution train location determination, independent of track circuits; continuous, high capacity, bidirectional train-to-wayside data communications; and train borne and wayside processors capable of implementing vital functions." There are many keywords in this definition, but we need primarily to focus on three: continuous, high capacity and bidirectional. Continuous means that the communication between train and wayside could be everywhere, not only at certain points; high capacity obviously means that the system is able to handle all the data needed with some restraints (delay, jitter, data rate, etc.); and finally, bidirectional implies that not only the train will receive data from the wayside but the wayside will also receive from the train. These three aspects are essential to understand how a CBTC system works.

To explain CBTC properly an entire book like this would be required, so we use a simplified model of it, with two basic use cases. In our model, the train is able to know (in a very precise and reliable way) its speed and its position (by reading beacons located on the track and using odometry techniques). The measured speed and position needs to be transmitted to the wayside equipment, which is responsible to tell every train how far it can go and how fast (this is called the 'authority movement'). So, the train transmits its position and speed, and receives the authority movement. Given that this information is critical, it needs to be delivered in a very short time, otherwise it would be useless. If a train experiences some kind of difficulty and finds itself unable to receive authorities of movement or to transmit its position, it will stop before going into a dangerous state (i.e., the risk of crashing into the next train).

Once we have this CBTC simplified model we start to look at the requirements to the communication subsystem. As the data to be carried is critical, we want a robust communication system, able to work even under the eventual failure of some elements. This implies a redundant solution for the communication network, with separate paths for the wirings, duplicated base stations, switches, routers, power supplies, etc. Regarding the wireless part, this approach is called 'double-layer' coverage, and it is a very expensive solution, as it represents (more or less) twice the cost that you would have on a 'single-layer' coverage. **Figure 1** is a picture of a double-layer deployment, with coverage by two different base stations at every single point of the track. Besides, this network also needs to be able to handle the bidirectional communication of the wayside with every train. Usually this communication between the wayside and the trains is done through a polling routine, where the wayside iterates from the first to the last train, requesting them to communicate their positions/speeds and sending to them the calculated authority of movement. This polling routine impacts on the whole performance of the system (and limits the number of trains a single wayside equipment can handle).

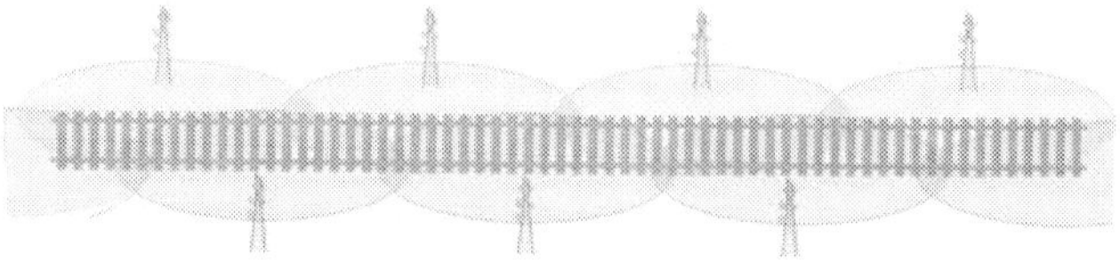

Figure 1. Double coverage deployment.

As the data volume to be sent is not very large (64–128 bytes per message every 1–2 s, depending on the supplier) and the required delay is less than 800 ms the quantitative requirements for the network are not very demanding. Therefore, all the aggregated data rate usually does not exceed 10–50 kbps, a figure that can be reached by almost every state-of-the-art wireless system (and even by some not so modern). In Section 3 we discuss in more depth the available wireless technologies, but since its early years, almost every commissioned CBTC has adopted IEEE 802.11 [4] technologies for its radio subsystem. These solutions are deployed in unlicensed (Industrial, Scientific and Medical, ISM) bands, something that has security implications. Ultimately, the trend for some vendors is to move instead to long-term evolution (LTE), but these cases are still an exception. For further details on the discussion see [5].

There are only a few CBTC vendors with a tested solution: Alstom, Ansaldo STS (now part of Hitachi), Bombardier, Siemens and Thales. There are many other signaling companies working on this technology, but the homologation procedures are very arduous and the operators frequently requesting to prove a tangible previous experience, this is an entrance barrier for new players. CBTC is not very much standardized (at least not like its high-speed counterpart ERTMS [European Rail Traffic Management System]), which means that vendors have a lot of liberty in the implementation of their solutions and CBTC not being interoperable, unfortunately for operators, with CBTC lines as they usually find out that their trains are captive on these lines.

The second group of mission critical services is the public safety (PS) issue, sometimes called 'professional mobile radio' (PMR). These systems are the evolution of the analog radiotelephony systems that were put into service in the early 80s. Strictly speaking, they are not safety systems (they do not protect the train like the signaling systems) but most operators and stakeholders act as if they were. As the OCC loses the communication with the train driver, some operators react to a breakdown of the radio telephony by evicting the train. The most common functionalities are the voice calls between the OCC and one or more trains. The addressing of these calls is usually required to be not only direct but also "functional" (based on the train number that may change every day) or "regional" (based on the location of the train).

There are two major technologies for PS radios: the first one is GSM-R (global system for mobile communications – railway), which carries PS data and signaling data in high-speed lines, mainlines (out of the scope of this book) and sometimes in commuter lines (like C4 in Madrid). The second technology is TETRA (terrestrial trunked radio) or any of its competitors. In this chapter we focus just on TETRA, as it is the most used PMR technology in the world. This is especially true for railways but its use is also widespread among police, firemen, border security, search and rescue, etc. All over the world it is assumed that the future of PMR systems is LTE (especially TETRA, whose association has publicly supported the effort of 3GPP LTE to provide PS functionalities [6]). In the next section, we discuss with more detail the feasibility of LTE for public safety.

The TETRA standard is very useful for railways because it allows operators, stakeholders, maintenance people, etc., to benefit from the following functionalities:

- Group calls: not just one-to-one calls, but also one-to-many.
- Functional addressing: i.e., the OCC wants to talk to the driver of train 2.
- Geographical addressing: i.e., the OCC wants to transmit a message to all trains that are in a certain area.
- Preemption: In the case of emergency, calls with less priority can be preempted in order to release resources for the higher priority ones.
- Direct communication: in the absence of a fixed network, two terminals can communicate with each other directly.
- High security levels: end-to-end encryption, authentication, integrity, etc.
- Data: although TETRA is not a broadband system, it is also able to transport short pieces of data from the train to wayside and vice versa.
- Push-to-talk (PTT): calls should be established in a very short time, usually, less than 500 ms. This represents a very demanding requirement for a network.

There are many TETRA vendors around the world, but it is a niche technology if we compare it with public mobile solutions such as LTE or UMTS (universal mobile telecommunications system).

2.2. Operator-oriented services

There are many operational services, the most common being CCTV (both real-time and recordings), passenger information systems (PISs) and telemetry of the train.

Passenger information systems deliver multimedia messages announcing passengers the proximity of the next station, the end of the line, etc. Most of the time only text messages are shown on a LED display at the end of the vehicle and/or audio recordings are played. However, modern subways and trams also provide video content on screens with information about nonoperational issues and also, very often, advertisement. This type of PIS is often called 'infotainment'. Some operators consider this system that provides a 'customer-oriented service' and not an operational one.

To provide an efficient passenger information service, the onboard system needs to know the location of the train with a certain grade of accuracy. This is done usually by integrating the PIS with the onboard signaling subsystem, deploying beacons in the track, or integrating the onboard PIS with a wayside system that knows the position of the train. The first method is the most accurate, it does not need a wireless system at least for the PIS (data remains in the train) but can be very expensive, because this integration can be very time consuming hard to put into service (many use cases and degraded modes). The second method is also costly, because beacons need to be installed and maintained, and sometimes it is prone to failures (a train misses a beacon). The third method can be less expensive if you already have a train-to-wayside radio, but this 'general purpose' radios are not as reliable as the PIS should be. Also, the integration difficulties remain. Of course, onboard infotainment systems need to download the multimedia contents to be shown and a broadband train-to-wayside radio is required.

Closed-circuit television (CCTV) is an operator-oriented service already available in almost every subway and tram of the world. Sometimes the recordings are kept in the train (if there is no train-to-wayside system available or there is no OCC integration to manage the video). Here we focus on the more general case where the video is available in the wayside and is also integrated with the train control management system (TCMS) of the train. There are two basic functions on CCTV systems for railways: real-time video and recordings. Real-time video can be watched in the cabin of the train by the driver and/or in the OCC. Some operators/stakeholders have 'security centers' where all the security aspects of the system are centralized (access controls, CCTV, security staff management, etc.) but in most cases this is part of the OCC. As real-time CCTV (video streaming) systems are very demanding in terms of resources, the train-to-wayside wireless system shall be properly designed. The other major function is the viewing and downloading of video recordings. Some operators force their onboard CCTV systems to download their recordings to be stored somewhere in the wayside (in a NAS [network-attached storage], probably), while others prefer to store it on the train itself and access them on demand through the wireless system.

Apart from the previous two functions, there are many other railway-related than usually enriching a CCTV system. Having the video of the upcoming platform in the cabin of the train can be very useful in the case of crowded stations; and integrating the smoke detectors with the onboard CCTV (via TCMS) so when a smoke alarm is triggered, the driver automatically can watch the nearest camera to the alarm/source of the smoke, are some examples. In both cases, a big effort on systems' integration shall be done, so it is important to keep in mind that a railway-CCTV system is not only a CCTV system placed in the railway system. CCTV systems, besides being relevant for security purposes, could also be helpful for operational ones (i.e., for driverless trains, cameras placed at the front of the train pointing to the track, etc.).

Finally, there are telemetry services where some data is transmitted out of the train to some/a particular wayside system, in order to analyze them, take decisions based on them, etc. There are two main final users of this data: train maintainers and railway operators. All these data needs to be obtained inside the train (sensorized), transmitted out of the train (usually through a gateway from the sensors network to the IP network) and finally, handled at the wayside (and maybe even stored on a database and integrated into the OCC or maintenance tools). In modern trains, the digital buses of the trains contain a lot of data of almost every single piece of the train, so it is rarely needed to deploy new sensors. This type of buses, deployed for TCMS purposes, usually follow the IEC 61375 [7] family of standards. In Section 5 we will know about the evolution of these standards.

2.3. Customer-oriented services

The last group of services is focused only on the client experience, with no safety implications or added-value for the operators. The most important service in this category is the Internet access for passengers within the train. At first sight it might seem that it is not a very challenging service to be provided, but many difficulties may appear:

- It is an extreme environment for electronic devices, with vibrations, high temperatures, electromagnetic interference (EMI), etc. These perturbations are usually handled together requesting that every onboard device complies with the railway normative.
- Vehicle penetration loss (the losses of the signal strength due to the presence of the vehicle) is usually between 15 dB and 25 dB, depending on the frequency of the signal and type of vehicle. This can be avoided using mobile relays.
- Cybersecurity. Having user data and safety data in the same network implies handling various security issues, much more complicated than having only safety and operational data in the network.

3. Networks and technologies

In this section we discuss the technologies that lie between the services explained in Section 2 and the physical layer to be covered later. Independently from the technology, any train-to-wayside system carrying data from different services should have some QoS (Quality of Service) policy, in order to implement prioritization schemes. This is not a trivial task (especially among legacy technologies). Security issues shall be addressed too, more than ever if the network is reachable for passengers (WiFi services, for example). Furthermore, to implement a successful train-to-wayside system it needs to be properly integrated in the ground network. Again, this is out of the scope of this book, but many networking issues may appear if both technologies are not compatible. For instance, if base stations need layer 2 connectivity, but this is not possible with the topology of the network, the performance of the train-to-wayside system will decay.

We cover two major technologies the 3GPP LTE [8] and the IEEE 802.11 [4] family of standards with a spin-off of the latter, but with proprietary modifications (this is, out of the open standard).

3.1. LTE

Undoubtedly, 3GPP LTE (long-term evolution) has a key importance in railways. Also, due to its popularity in the public mobile world, it is also very frequent to find out that passengers have LTE coverage. Furthermore, the intention of the 3GPP (the world organization responsible of the standardization of LTE) is to include railway use cases and functionalities. However, the 3GPP specifications for LTE are a living thing, with frequent releases and with the vendors pushing to develop their solutions. This implies that what is written here can be outdated in a few years, so our suggestion is to learn here the basics of LTE, the philosophy behind it while also trying to keep up with all the novelties that the market could offer.

The first problem that appears when you consider LTE as a candidate for your wireless system is the spectrum. In radio communications you cannot transmit anything you want due to the regulation that you shall comply with. So, first of all, you need to know in which band your system is going to work. There are two possible types of LTE networks: public (a service typically provided by mobile operators like O2 or Vodafone), where the railway users compete with the passengers and everybody else for the resources; or private, where the railway services are the unique users of the network. An arrangement between both extreme options is also possible (for example, a public mobile operator being a piece of their spectrum for railway use, de facto creating a 'private LTE') or, simpler, the railway service being one plain customer of the mobile operator.

The LTE architecture is very flat, having only a core and the access network (see **Figure 2**). The access network consists of a set of base stations called eNB ('evolved node B') which handles the radio interface between the user equipment (UE) or terminals, on one side, and the interface with the core, on the other side. There are many possible implementations for the eNB, depending on the scenario, for example, one eNB, two radio heads and a microcore all in one box; in the other side, in a large area, many eNB connected through a dense fiber network to a centralized core. Therefore, flexibility is one of the key features of LTE.

The functions of the eNB belong to two different sublayers [10]: RRM (radio resource management) and MAC (media access control). First, RRM functions are user admission control, scheduling (dynamic assignment of resources to users), interference control between eNB to avoid interferences, mobility management (handovers) and many other minor functions.

The second sublayer media access control (MAC), with two main differences between the uplink (from the UE to the eNB) and the downlink (eNB to the UE). In the downlink, LTE uses OFDMA (orthogonal frequency division media access), a technique that separates the transmissions destined to different users by assigning them different blocks of data subcarriers. Given that every eNB manages limited resources in both time and frequency (the so-called bidimensional "grid" of time slots and subcarriers, see **Figure 3**), this technique gives eNB enough flexibility to manage their pool of transmissions. In the uplink, due to limitations in the dynamic range of the amplifiers, SC-FDMA ('single-carrier FDMA'), which is a variation of OFDMA, is used instead. Another example of the flexibility of LTE is that the total bandwidth can be 1.4, 3, 5, 10, 15 and 20 MHz. LTE can work in paired mode (also called FDD mode with a band for the uplink and another one for the downlink, both with the same width) or

http://dx.doi.org/10.5772/65585

unpaired mode (TDD mode, with only one band for both up and downlink) [10]. Carrier aggregation can be used in the recent versions of LTE known as LTE-Advanced [10] and LTE Advanced Pro [9], combining two or more radio frequency (RF) channels of the same, or different, frequency bands to provide very high data rates.

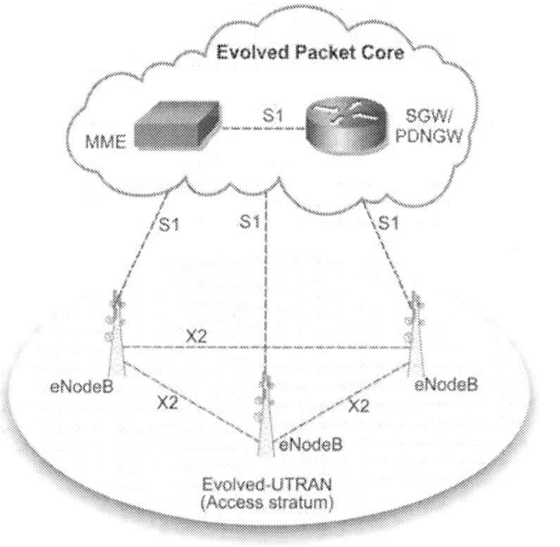

Figure 2. LTE architecture, depicting both access stratum and core.

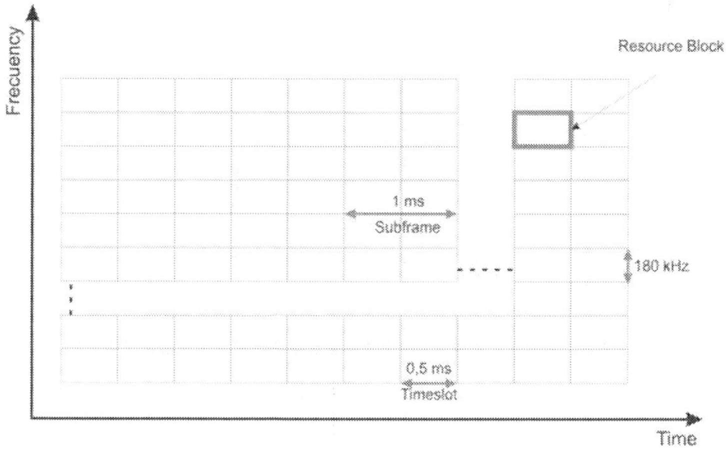

Figure 3. Radio resources on a LTE eNB.

The spectrum problem

An issue that shall be addressed when working with wireless is the spectrum allocation. Every single country in the world regulates the use of it, so we are not at liberty to transmit wherever we want. There are three main possibilities for this allocation:

Unlicensed bands: In these bands no license from the regulator is required. You only need to check that the emitted power does not exceed the maximum allowed, but as anybody can transmit, you need to take care of the interferences that they may cause. There are many unlicensed bands all over the spectrum; the most popular (for both railway and nonrailway use) are the 2.4 GHz and the 5 GHz.

Licensed bands for railway use: In some cases, the regulators allocate a piece of the spectrum for railway applications, but, unfortunately, this is not very common—for instance, the GSM-R band in Europe (873–880 MHz for the uplink; 918–925 MHz for the downlink).

Licensed bands for non-railway use: These bands represent the majority of the spectrum. No railway applications can use this band, unless an agreement with the owner of the rights is signed.

As mentioned in the previous section, when explaining public safety services, LTE is supposed to be the next dominant technology in this field. However, LTE as it is now is not able to provide the PS required functions [6] (future 3GPP LTE releases will come to solve this problem).

Also in 3GPP LTE's roadmap [6] is planned to include a very useful technology for railway use: the mobile relay (MR) [11]. It is an onboard device that divides into two parts, the link between the UE and the eNB: one part between the UE and the MR (inside the train), and another one between the MR and the eNB (out of the train). This approach offers many advantages like reducing losses caused by the structure of the vehicle, the possibility of having better DSP (digital signal processing) techniques than in cheaper UEs to avoid multipath, Doppler and other undesired effects; group handovers can be performed (only the MR does the handover); obviously, it is an opportunity gap for railway operators to increase their incomes through a partnership with a mobile operator, etc. As usual, there are some drawbacks, because the MR increases the end-to-end latency (it implies one more hop); as every handover, it can fail resulting on a massive communication loss (every UE attached to the MR goes down if the eNB–MR link goes down); it shall be integrated with the TCMS of the train, which implies ad hoc designs and extra costs. However, as we write this text, no LTE vendor has on its portfolio anything like a MR.

3.2. IEEE 802.11

The IEEE 802.11 standard, or family of standards, is a group of requirements only affecting the physical and the MAC layers. It has many amendments, which will be briefly explained later, and it has gained widespread recognition because the popular 'WiFi' is based on this standard.

802.11 is a half-duplex technology, which means that both sides of the communication channel can transmit, although not at the same time.

In railways this technology is present in two different ways: to deploy private networks for safety services (like CBTC) and operator-oriented services, and also to provide access to the Internet for passengers (i.e., a public network in stations and inside the trains).

It is deployed in unlicensed bands (ISM bands) at 2.4 GHz and/or 5 GHz, depending on the amendment of the standard 802.11. These ISM bands are usually very crowded, with many APs interfering. This is the great challenge here, especially in the 2.4 GHz, which is far more crowded than the 5 GHz one. However, 802.11 technologies are very resilient and they are able to work even in noisy environments. In **Figure 4**, it can be observed that from the 13 possible channels at 2.4 GHz only three are nonoverlapping (typically channels 1, 6 and 11 are used). In the 5 GHz band the situation is different because there are more nonoverlapping channels (see **Figure 5**) and we can also aggregate bandwidths in order to form 40-, 80- and 160-MHz wide (and the 20-MHz ones, too). Please note that not every channel is available everywhere, because it depends on the country where you are in.

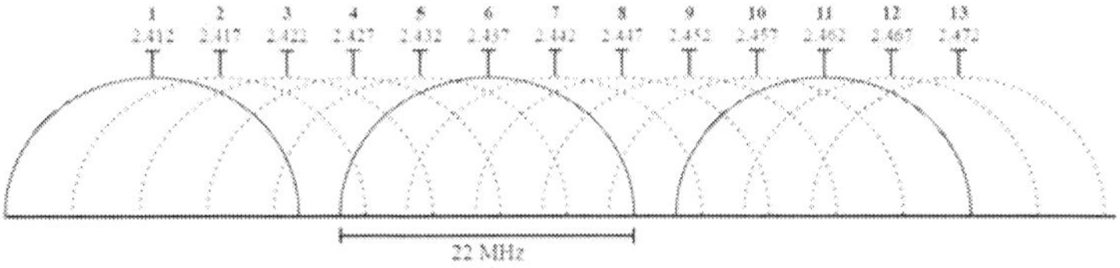

Figure 4. 2.4 GHz channels for IEEE 802.11b/g/n.

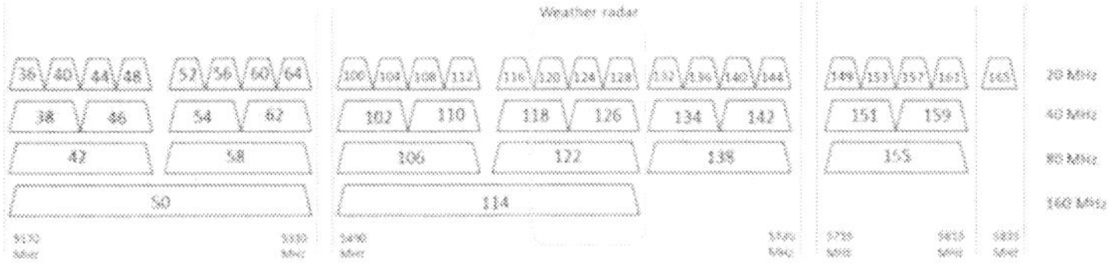

Figure 5. 5 GHz channels for IEEE 802.11.

Since 1997 there have been a lot of amendments to the baseline IEEE 802.11 specification. Here we only focus on the most popular for any application (and also railway-related): a/b/g/n/ac. From a practical point of view, amendments a/b/g/n/ac are standards on its own, so it is very common to specify an IEEE 802.11 technology only using the amendment identifier.

The architecture of this technology is similar in many aspects to LTE. We have user devices (called terminals or clients), base stations (access points, APs) and also controllers for the base stations. On IEEE 802.11 only the protocol between a client and the AP is standard. For example, communication between APs and controllers (and also between controllers) does not need to be standard, so vendors have developed their own noninteroperable solutions.

Amendment 'a' was the first of a long list and it had the same core protocol as the baseline IEEE 802.11, but introduced OFDM (orthogonal frequency-division multiplexing, with 52 subcarriers) and has a nominal data rate of up to 54 Mbps. As they are very dependent on the environment it is difficult to provide practical or average data rates. Unlike next amendments, the 'a' one works at 5 GHz. Amendment 'b' is also an evolution of the baseline IEEE 802.11 with some features in common with the latter (2.4 GHz band, direct-sequence spread spectrum [DSSS] modulation) while others not ('b' achieves nominal data rates of up to 11 Mbps and 'baseline' only 2 Mbps). This technology was the first IEEE 802.11 which was mainstream popular and also was implemented in some CBTC systems (i.e., Bombardier's Cityflo 650) plus other general-purpose train-to-ground systems. The next step in the IEEE 802.11 evolution is the 'g' amendment, which is similar to 'a' but works in the 2.4 GHz band. IEEE 802.11n was the turning point for the IEEE 802.11: able to work in both the 2.4 and 5 GHz bands, it reaches higher data rates by using wider channels (40 MHz and also 20 MHz) and multiple input, multiple output (MIMO). The results are nominal data rates of 144 Mbps for 40 MHz-wide channels and 72 Mbps for 20-MHz ones. The last amendment covered here is the 'ac' one. It is disruptive with the earlier ones in the sense that it only works in 5 GHz and allows multiuser MIMO, channels of 20, 40, 80 and 160 MHz (or 80 + 80 MHz) wide, which leads to nominal data rates of 750 Mbps. Early tests of train-to-ground systems based on IEEE 802.11n and 'ac' proved data rates of 50 Mbps and 300 Mbps, respectively.

IEEE 802.11 is very frequently used in CBTC systems and operator-oriented services for train-to-ground communication. Obviously, 'WiFi' services for passengers in train stations, and also recently inside trains, implement this technology. Inside the trains there is also the need to backhaul the WiFi data from passengers out of the train with a train-to-ground link (which could be IEEE 802.11, LTE or a proprietary development). Some railway operators also use this technology to provide connectivity to two coupled consists, but sometimes the 'train inauguration problem' (discovery of the train topology to establish and maintain the train network) requires proprietary algorithms.

3.3. Proprietary solutions

IEEE 802.11 standard works in a very efficient way in static environments but in nomadic scenarios as well (punctual movements, at low speed, typical in walking areas like offices, homes or malls). But this is not the case for train-to-ground links where the train moves fast (subways up to 140 km/h). If a client wants to move to another AP area without losing connectivity, one AP needs to transfer the user to the other. This is generally called 'handover' and sometimes (mostly in WiFi context) 'roaming'. This procedure is critical for the performance of the whole network and IEEE 802.11 solutions suffer when performing it, as they are not optimized for high mobility scenarios like subway lines or tramways. So, some modifications need to be done in the standard to make this roaming more efficient and fast. These modifications neglect the interoperability between the 'modified' solution and the standard one. Of course, handover optimization is not the only reason for performing

http://dx.doi.org/10.5772/65585

proprietary modifications to standard technologies. Security techniques, train inauguration algorithms, routing protocols, mesh architectures, etc. are sometimes the cause of these modifications.

Some examples of proprietary technologies for railways are Fluidity (developed by Fluidmesh), Traincom (by Telefunken, now Siemens), etc.

Railway certification

Something very frequently missed when installing electronic equipment onboard is the importance of the railway certifications. Onboard equipment is exposed to many issues that threaten seriously their performance and reliability. The reason behind this phenomenon is that temperature usually ranges from −40°C to +85°C; due to the presence of high voltages and currents the electromagnetic fields are very large; trains also experience heavy vibrations, etc. Moreover, this equipment should be ready to work 24/7 for 20 years at least (trains are usually designed to last 30 years). Unless we address these threats making the onboard equipment more resilient, the impact on the performance would certainly decay. This means that no electronic device shall be installed on any train without the proper certification. The most common is the EN50155 [12], which references also almost 50 other norms. It covers temperature, power supply, humidity, vibration, shock, and electro magnetic compatibility (EMC) issues.

---

4. Physical layer

In this section we cover propagation and antenna issues. Propagation models are very much needed because knowing how the signal is going to behave is mandatory in order to design communication systems. From a system's point of view it is not usually needed to know models in a very detailed way, but it is important to understand how the signal propagates, some basic principles of antenna design and how to operate with the models.

In general terms, the layer 1 (physical layer) design of a wireless system starts with some premises (frequency, maximum power, minimum signal in the receiver, desired overlap between cells, etc.) and then you place the base stations. Minimum signal strength is a parameter that is heavily dependent on the service. For instance, VoIP (Voice over IP) deployments require stronger signals than CBTC ones, as each service has its own KPIs (key performance indicators) and demands different QoS from the network. It is out of the scope of this chapter to provide precise figures about these values, because sometimes they are vendor-dependent (CBTC, for example) and sometimes they depend on subjective criteria (i.e., VoIP). Here we present some models to calculate the received power (given the transmitted power) to perform such a layer 1 design.

4.1. Urban scenario

Regarding urban transport we need to cover the so-called urban propagation models. These models tell us how signals behave between the transmitter and the receiver. Tramways usually run in classic urban environment, that is, in surface areas surrounded by buildings. The modelization of these environments was started early in the sixties in Japan by Okumura and perfected in the next decade by Hata. The result is the famous Okumura-Hata model [13], where the path loss is:

$$PL(dB) = 69.55 + 26.26\log(f) - 13.82\log(h_t) - a(h_r) + (44.9 - 6.55\log(h_t))\log(d) \quad (1)$$

where f is the carrier frequency in MHz, h_t and h_r denote the height in meters of transmitter and receiver, respectively, and d is the distance between them in kilometers. This model works well if the frequency is below 1.5 GHz, so for higher ones we shall use a different path loss expression. This is known as the COST 231 extension to Hata model [14]:

$$PL(dB) = 46.3 + 33.9\log(f) - 13.82\log(h_t) - a(h_r) + (44.9 - 6.55\log(h_t))\log(d) + C_M \quad (2)$$

where C_M is 0 dB for medium cities and suburb areas and 3 dB for dense metropolitan areas. There are other limitations in the feasibility of these two models as they are intended for larger cells (typical in early mobile deployments) but it is still a good entry point to the design of a tram-to-wayside wireless system. However, the main limitation is that they do not work if the frequency is higher than 2 GHz. Working in higher bands require another models, like Winner II [15] which is specified for the band between 2 and 6 GHz.

4.2. Characterization of the tunnel scenario

The other environment where urban rail operates is the tunnel. The modeling of the propagation in tunnels is a task that is not trivial at all. There are some effects that make this characterization very difficult to accomplish and it is quite different from the general model.

As mentioned before, railways are a hostile environment for electronic devices in general. In the particular case of radio communications in tunnels, we need to add some extra problems that make communication even harder: presence of large metallic masses in movement (trains), arbitrary tunnel cross-sections, frequent changes on these sections, obstacles like cabinets, tracks, catenaries, etc.

To properly model this behavior, there are three types of models [16] based on modal-theory [17], general theory of diffraction (GTD) [18] and hybrid approaches [19]. Modal theory considers the tunnel an oversized waveguide, so in order to let signals to propagate the frequency it needs to be higher than the cut-off frequency of the tunnel [17]. The propagating signal is composed of the aggregation of many modes, each one of them with its own attenuation and phase constant. Another limitation is that properly modeling some details of the environment (a cabinet, the track, a pit, etc.) can be very arduous [16].

To overcome some of the limitations of modal-based models we use ray-tracing techniques [18], or more generally-speaking, those based on the GTD. These models consist on the launching of a set of rays calculating the reflections and diffractions of each ray until they get to the receiver. To be done properly, significant computing resources are required, as well as a proper characterization of the surfaces of the elements in the tunnel. A good ray tracing-based tool is explained in [20].

The third group of models is the hybrid ones, which combine the advantages of the previous methods plus some corrections provided by measurements in the field. This is a hybrid approach of stochastic and deterministic models. Assuming that no model is perfect and site surveys are always needed, here we highlight an empirical method that, in our opinion, is accurate and easy enough to handle. It assumes that every tunnel is circular (if not, a correction term 'κ' is introduced for other shapes) and estimates the losses α (dB/m) as [16]:

$$\alpha = \kappa\lambda^2 \left[\frac{\varepsilon_r}{a^3\sqrt{\varepsilon_r - 1}} + \frac{1}{b^3\sqrt{\varepsilon_r - 1}} \right] \tag{3}$$

where κ is the shaping parameter, which varies with the cross-section of the tunnel (5.09 for circular; 4.34 rectangular; 5.13 arched), λ is the wavelength, ε_r the dielectric permittivity of tunnel walls and a and b are the width and the height of the tunnel, respectively. This estimation is very useful for physical layer designs in tunnel but its accuracy suffers when curves are present in the track.

To conclude this section we shall also refer to a scenario that is very much used in tunnels: the leaky feeder or leaky cable. This feeder is an alternative to discrete antennas especially useful in complex environments like winding corridors, staircases, castles and tunnels. It consists of a coax wire with many slots on its outer jacket that let the energy inside the wire to get out of it. The receiving process is the opposite, with signals getting into the slots, traveling across the leaky coax and reaching the receiver. The main benefit of this approach is to have a uniform coverage and that we do not need to perform complex calculations to estimate the coverage in the tunnel. It is true that this argument was usually given before the previous accurate models were developed, but it is still valid today in many subways. The main drawback is the high cost that has the installation of the leaky coax in the tunnel wall. The leaky feeder approach is usually followed when frequencies are under 2 GHz, but it is also possible to find leaky feeders that may work at higher frequencies.

4.3. Other scenarios: inside the train and vehicle-to-vehicle

Besides from the train-to-wayside, it is also important to emphasize other scenarios for wireless communication in urban rail: inside the vehicle (typically, for sensor networks or WiFi deployments); vehicle-to-vehicle (for the train backbone, replacing the Scharffenberger couplers); and also train-to-train (for future signaling systems). With the sole exception of the first one, the modeling of these channels is still on a very early stage of development. There are some research papers [21] concerning this topic and some ongoing projects which will be introduced in the next section.

4.4. Antennas

Antenna engineering is a huge field that exceeds by much the scope of this chapter, but it is highly advisable to understand properly its basics. Here we cover two types of antennas: the wayside and the onboard. There are two main issues regarding both of them that require to be addressed: choosing the right antenna and placing them in the right location. Installing an antenna on a tunnel is usually arduous due to the gauge limitations. Onboard antennas tend to be omnidirectional and the installation tends to be duplicated (two cabins instead of just one) increasing the performance of the system, especially in tunnels, where the train blocks the signal when the wayside antenna is in the opposite side. Wayside antennas are directive in most of the cases but not always. It is important to highlight that some installations may require more than one antenna on a single spot. This may be needed in diversity, beamforming or MIMO setups. In these cases, the typical spacing between antennas is λ (one wavelength) or $\lambda/2$.

5. The future of wireless communication systems for urban rail: 5G and other research directions

There are many initiatives around the world that will influence very much the future of wireless systems for urban rail. In the final section, we will describe the most relevant lines of research actually in progress. The structure is the same as in the entire chapter: services, technologies and physical layer issues.

There are two main triggers for this research: major projects, with total or partial public funding and the participation of vendors, integrators, railway operators, universities, etc. on one side, and all the investigation that is privately done by companies. Obviously, it is more difficult to know any progress coming from the latter, so we will focus on the former. In the last five years, the European Union was developing numerous research projects on wireless aspects of railways. Some good examples are Roll2Rail, Shift2Rail, SECRET, Systuf, Tecrail, and Integrail, but there are many more. Roll2Rail and Shift2Rail are more initiatives than projects, acting as 'containers' for projects on many railway-related topics (for example, only work package 2 of Roll2Rail is focused on wireless, but it is still a very large project). Most of these trends are very unlikely to be implemented in real in-operation systems in the short term.

5.1. Upcoming services

Once you have a CBTC system on a subway line it could help the operator to provide many other services. Some good examples are passenger information (after all, the system that knows better the location of the train is the CBTC) and many other operator-related ones. It is very difficult to determine how CBTC will evolve (especially as vendors are very secretive concerning their research), but we can predict that one characteristic that could be explored is the direct communication between trains (in addition to the train-to-wayside communication already available). If we introduce direct communications between trains, a train could transmit its position and speed to the precedent one (see **Figure 6**), which could calculate its authority of movement using this information and also its own position and speed. This is the same as in the conventional CBTC but it could be performed in shorter times, (end-to-end delay decreases). This concept needs a reliable on-board device-to-device communication link, which sometimes could be not available (when trains are not very close or are some obstacles between them). As far as we know, the pioneer here is Alstom, with its Urbalis Fluence technology, still under test.

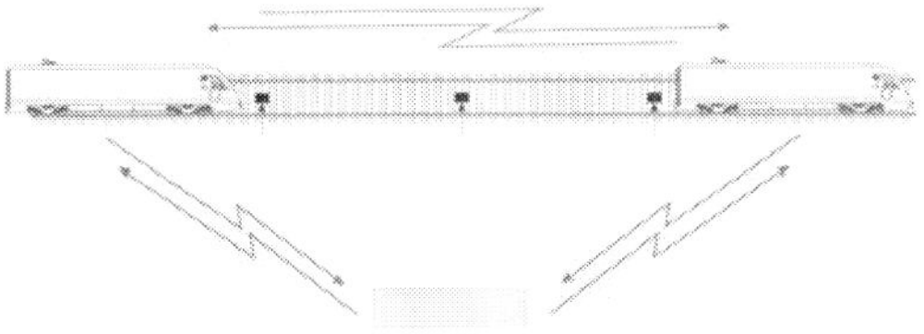

Figure 6. Schematic representation of a CBTC system with V2V (vehicle-to-vehicle) capabilities (in blue) plus the 'traditional' V2I (in green).

Another service that is gaining importance is the sensorization of onboard systems, transmission to wayside servers in order to be analyzed for operational and maintenance purposes. This methodology is very much aligned with the Internet of Things (IoT) paradigm as well as with 'Big Data'. It is still in progress in many subways and trams around the world, but is a very promising technology for operators and rolling stock maintainers, especially as it enables the opportunity to monitor, in real-time, the status of the entire fleet. This is also key for driverless trains. But many challenges lay ahead that need to be addressed. For example, onboard bus topologies are not in practice as standardized as they should be, with many legacy networks and buses still in use; data from the sensors do not reach the wayside; there are many security issues; the storage and CPU requirements in the wayside are enormous, especially in large subway networks with many trains. Despite of this, it is very likely that in the near future the IoT paradigm will be another history of success for urban rail.

Another trend that is being discussed at the time we write this chapter is the feasibility of a wireless train network. That is, replacing the wired buses carry all the TCMS data of the trains with wireless links. This is a move very similar to offices and homes, where many wires have been replaced by Bluetooth and WiFi links. However, the challenges are many: RAMS (reliability, availability, maintainability and safety), because railways are very demanding concerning this issue; security concerns too, due to the fact that the data is far more exposed in a wireless link; availability of a mature technology able to carry the data inside the train, between vehicles, etc. This is the main purpose of WP2 of the European Project Roll2Rail [2].

Public safety services are very likely to be provided in the future over an LTE network [6]. TETRA association has publicly supported this move [6] so it is only a matter of time. Huawei, a major LTE player, has strongly supported this view with their technology eLTE, which is able to provide many public safety features over a nonstandard LTE core.

Finally, another important challenge for the future is the integration of tramways on smart cars' platforms. All the work that it is being carried out on autonomous cars, smart highways, smart cities, etc. will also need to integrate tramways on it.

5.2. Future technologies

At the moment, the most promising wireless technologies for use in these environments are IEEE 802.11 and 3GPP LTE. Maybe they would require proprietary improvements, but it is clear that both of them are trying to be strong at railways use cases. In the case of IEEE 802.11, there are two main lines of research: one centered in the evolution of 'ac', which is the ongoing 'ax' and another one focused on the so-called 'WiFi Gigabit' which works on the 60 GHz band (that is, 802.11ay, the evolution of IEEE 802.11ad). The release of the first draft for both 'ax' and 'ay' standards is expected on 2016 and 2017, respectively. 60 GHz band is more likely to be deployed inside vehicles than for train-to-wayside or train-to-train scenarios.

As it was mentioned in Section 3, 3GPP LTE is introducing more railway-related use cases, like mobile relays. Besides from this standardization effort, vendors are very likely to introduce algorithms with optimizations for radio resource management tasks, especially handovers. This would happen when the 4G (Fourth Generation) coverage on high-speed lines gets denser. Another technology that could help the market penetration of LTE in railways is the unlicensed LTE (LTE-U). It is a technology in the roadmap for 3Q 2016 for the main LTE vendors. There are three different philosophies on LTE-U: (1) LTE on unlicensed bands but as a secondary best-effort carrier for user data (in the downlink only), remaining the licensed band as a primary carrier (for both user and control data). This is also known as LAA, license-assisted access. The second offload alternative for mobile operators is to send this traffic directly to WLAN (wireless LAN) networks (LTE–WiFi link aggregation or LWA). And the third one, "standalone LTE-U" does not require operator support or licensed bands to work.

5G (fifth generation) technologies are starting its standardization. There are many governments and organizations in China, Korea, Japan, and the European Union that already have their committees, where the railway industry is also very active with some research projects that include some railway-related use cases, like the Spanish Enabling5G. The first three generations of mobile communications turned their back on railways (with the remarkable exception of GSM-R); 4G started to look at railways, and the railway industry is really implied on 5G standardization.

Finally, an interesting debate is to dedicate some spectrum for railway use (excluding the 5 + 5 MHz of the GSM-R band, which is not available for urban rail). A joint task force composed by the ETSI (European Telecommunications Standards Institute) and the UITP (International Association of Public Transport) has recently launched a very interesting technical report [22] on this topic, where it is suggested to share part of the 5.9 GHz band (dedicated to ITS) with the CBTC systems (now on ISM bands, especially at 2.4 GHz). This is an interesting move for both organizations as it could open a door for a licensed band for railways.

5.3. Physical layer research trends

There are two main challenges: the research of better channel models and the development of more suitable antennas for railways. A recent paper [23] identifies the most relevant open issues on channel modeling in railway environments (modeling the influence of vehicles, not only for train-to-wayside channels but also for train-to-train, vehicle-to-vehicle and intravehicle). In the train-to-train channels the modeling of nonstationary scenarios is still an open issue [24]. In the antenna field, the most relevant aspect is how to design MIMO antennas for the onboard, with 4 × 4, 8 × 8 setups and also to integrate them appropriately in the car body shell.

6. Conclusions

Wireless communications has revealed as a key player in railways. It has improved both safety and security, provided a better quality of service to passengers, reduced operational costs, etc. In this chapter we have explained the most important urban rail services, technologies that make them possible and finally, the physical environment where the wireless communication happens. Wireless technologies tend to be more standard every day (3GPP LTE and IEEE 802.11 lead the way). We have seen why the physical layer is still complicated to handle, especially in tunnels. Finally, we have introduced the most relevant research lines in this field to let the reader know how the near future is likely to be.

Abbreviations

4G	fourth generation
5G	fifth generation
AP	access point
CBTC	communications-based train control
CCTV	closed-circuit television
DSP	digital signal processing
DSSS	direct-sequence spread spectrum
EMC	electromagnetic compatibility
EMI	electromagnetic interference
eNB	evolved Node B
ERTMS	European Rail Traffic Management System
ETSI	European Telecommunications Standards Institute
GSM-R	Global System for Mobile Communications – Railway
GTD	general theory of diffraction

IoT	Internet of Things
ISM	Industrial, Scientific and Medical
ITS	intelligent transportation systems
KPI	key performance indicator
LAA	license-assisted access
LTE	long-term evolution
LTE-U	LTE unlicensed
LWA	LTE–WiFi link aggregation
MAC	media access control
MIMO	multiple input, multiple output
MR	mobile relay
NAS	network-attached storage
O&M	operation & maintenance
OCC	operational control center
OFDM	orthogonal frequency-division multiplexing
OFDMA	orthogonal frequency-division multiple access
OSI	open systems interconnection
PIS	passenger information system
PMR	Personal Mobile Radio
PS	public safety
PTT	push-to-talk
QoS	quality of service
RAMS	reliability, availability, maintainability and safety
RF	radio frequency
RRM	radio resource management
SC-FDMA	single-carrier FDMA
TCMS	train control management system
TETRA	terrestrial trunked radio
UE	user equipment
UITP	International Association of Public Transport
UMTS	Universal Mobile Telecommunications System
V2V	vehicle-to-vehicle
VoIP	Voice Over IP
WLAN	wireless LAN

Author details

Juan Moreno García-Loygorri[1*], José Manuel Riera[2] and Carlos Rodríguez[1]

*Address all correspondence to: juan.moreno@metromadrid.es

1 Engineering and R & D Department, Metro de Madrid, Madrid, Spain

2 Departamento de Señales, Sistemas y Radiocomunicaciones, Universidad Politécnica de Madrid, Madrid, Spain

References

[1] Shift2Rail: driving innovation on railways [Internet]. Available from: http://ec.europa.eu/research/press/jti/factsheet_s2r-web.pdf [accessed: 29/JUL/2016]

[2] Roll2Rail Project [Internet]. Available from: http://www.roll2rail.eu

[3] IEEE Standard for Communications-Based Train Control (CBTC) Performance and Functional Requirements. IEEE Std 1474.1-2004. 2004.

[4] IEEE 802.11 Standards [Internet]. Available from: http://standards.ieee.org/about/get/802/802.11.html.

[5] J. Moreno, C. Rodríguez. Carrying CBTC commands over a public LTE network: a railway operator perspective. Eurotransport. 2015, pp. 77-80.

[6] R. Ferrus, O. Sallent, G. Baldini, L. Goratti. LTE: the technology driver for future public safety communications. IEEE Communications Magazine, vol. 51, issue 10, October 2013, pp. 154–161.

[7] International Electrotechnical Commission (IEC) 61375–1:2012. Electronic railway equipment—Train communication network (TCN)—Part 1: General architecture.

[8] 3GPP website. [online] http://www.3gpp.org/. Retrieved on October 15 2016.

[9] A. Bhamri, K. Hooli, T. Lunttila. Massive carrier aggregation in LTE-advanced pro: impact on uplink control information and corresponding enhancements, in IEEE Communications Magazine, vol. 54, no. 5, pp. 92–97, May 2016.

[10] E. Dahlman, S. Parkvall, J. Skold. 4G: LTE/LTE-Advanced for Mobile Broadband. Second edition, 2013.

[11] Lin Chen, Ying Huang, Feng Xie, Yin Gao, Li Chu, Haigang He, Yunfeng Li, Feng Liang, Yifei Yuan. Mobile relay in LTE-advanced systems. IEEE Communications Magazine, vol. 51, issue 11, November 2013, pp. 144–151.

[12] EN 50155:2007 CORR:2010. Railway applications—Electronic equipment used on rolling stock.

[13] M. Hata. Empirical formula for propagation loss in land mobile radio services. IEEE Transactions on Vehicular Technology, vol. VT-29, no. 3, 1981, pp. 317–325.

[14] L.M. Correia. A view of the COST 231-Bertoni-Ikegami model. 2009 3rd European Conference on Antennas and Propagation, Berlin, 2009, pp. 1681–1685.

[15] Wireless World Initiative New Radio, part II. Available: http://www.istwinner.org/deliverables.html.

[16] A. Hrovat, G. Kandus, T. Javornik. A survey of radio propagation modeling for tunnels. IEEE Communications Surveys & Tutorials, vol. 16, issue 2, Second Quarter 2014, pp. 658–669.

[17] D. Dudley, M. Lienard, S.F. Mahmoud, P. Degauque. Wireless propagation in tunnels. IEEE Antennas and Propagation Magazine, vol. 49, no. 2, pp. 11–26, April 2007.

[18] C.F. Yang, B.C. Wu, C.J. Ko. A ray-tracing method for modelling indoor wave propagation and penetration. IEEE Transactions Antennas and Propagation, vol. 46, no. 6, pp. 907–919, 1998.

[19] Y. Wang, S. Safavi-Naeini, S. Chaudhuri. A hybrid technique based on combining ray tracing and FDTD methods for site-specific modeling of indoor radio wave propagation. IEEE Transactions Antennas and Propagation, vol. 48, no. 5, pp. 743–754, 2000.

[20] C. Briso, C.F. López, J.R.O. Fernández, S. Pérez, D. Draskovic, J. Calle-Sánchez, M. Molina-García, J.I. Alonso, C. Rodríguez, C. Hernández, J. Moreno, J. Rodríguez-Piñeiro, J.A. García-Naya, A. Castedo, A. Fernández-Durán. Broadband access in complex environments: LTE on railway. IEICE Transactions on Communications, volume E97-B, no. 8, pp. 1514-1527.

[21] P. Unterhuber, S. Pfletschinger, S. Sand, M. Soliman, T. Jost, A. Arriola, I. Val, C. Cruces, J. Moreno, J.P. García-Nieto, C. Rodríguez, M. Berbineau, E. Echeverría, I. Baz. A survey of channel measurements and models for current and future railway communication systems. Mobile Information Systems, 2016.

[22] Railways Telecommunications (RT); Shared use of spectrum between communication based train control (CBTC) and ITS applications. ETSI TR 103 442 V1.1.1 (2016–05) p. 1–26. [online]: http://www.etsi.org/deliver/etsi_tr/103400_103499/103442/01.01.01_60/tr_103442v010101p.pdf.

[23] Bo Ai, Xiang Cheng, Thomas Kürner, Zhang-Dui Zhong, Ke Guan, Rui-Si He, Lei Xiong, D.W. Matolak, D.G. Michelson, C. Briso-Rodríguez. Challenges toward wireless communications for high-speed railway. IEEE Transactions on Intelligent Transportation Systems, vol. 15, issue 5, October 2014, pp. 2143–2158.

[24] J. Moreno, J.M. Riera, L. de Haro, C. Rodríguez. A survey on the future railway radio communications services: challenges and opportunities. IEEE Communications Magazine, October 2015, pp. 62–68.

Chapter 4

Airborne and Ground-Borne Noise and Vibration from Urban Rail Transit Systems

Konstantinos Vogiatzis and Georges Kouroussis

Additional information is available at the end of the chapter

http://dx.doi.org/10.5772/66571

Abstract

The environmental effect of ground-borne vibration and noise generated by urban rail transit systems is a growing concern in urban areas. This chapter reviews, synthesizes and benchmarks new understandings related to railway vibration and associated airborne and ground-borne noise. The aim is to provide new thinking on how to predict noise and vibration levels from numerical modelling and from readily available conventional site investigation data. Recent results from some European metropoles (Brussels, Athens, etc.) are used to illustrate the dynamic effect of urban railway vehicles. It is also proved that train type and the contact conditions at the wheel/rail interface can be influential in the generation of vibration. The use of noise-mapping-based results offers an efficient and rapid way to evaluate mitigation measures in a large scale regarding the noise exposure generated to dense urban railway traffic. It is hoped that this information may provide assistance to future researchers attempting to simulate railway vehicle vibration and noise.

Keywords: structural vibration, railway vibration, environmental noise, vibration assessment, measurement, standards, human effect, building simulation, noise mapping, LRT

1. Introduction

The rapid population growth and its concentration in urban and metropolitan areas are creating new challenges and demands to mobility. The development of railway networks is therefore unavoidable and comprises the construction of new networks and/or the extension of existing ones. This increase in mobility often causes issues, notably noise and vibration which are important and must be managed for the coming decades [1]. Indeed, this problem is of growing importance: vibratory nuisance affects not only people (comfort and health) but also buildings (cracks as main and initial damage [2]). A recent example comes from Brussels (Belgium), which possesses a dense railway network in the city consisting of urban tramways, underground metros and regional trains and which receives numerous complaints: more than 280 complaints concerning the T2000 tram have been noted to the present day [3]. Based on this observation, comprehensive and integrated approaches are required to gain new powerful insights. Considerable efforts have been made in order to reduce the generated vibrations in the vehicle, improving the passengers' comfort, but the ground vibration problem must also be solved. In a large number of situations, the influence of vibration on structural damage in buildings and on people inside buildings can no longer be neglected. Switzerland's national railway company estimated that 1200 million euros were required to fix vibration problems across the country's network [4]. There has recently been a global development in rail infrastructure, which appears set to continue. This growth was associated to ground vibration awareness and triggering more severe standards regarding limits not to be exceeded. An active research area became evident by placing extensive efforts on predicting vibration levels with increased accuracy. This allows understanding more faithfully human perception of vibration.

The vibration of the building structure close to urban railway generates ground-borne noise, which can cause disturbance to the occupants. Sleep disturbance and annoyance, mostly related to transportation noise, comprise the main burden of environmental noise.

The mechanism of rolling noise is well mastered by the researchers in this field. The dominant source of noise is the rolling noise, generated by the interaction of rotating wheels and the rail. Both structures vibrate and radiate noise through vibro-acoustic effects. Similar mechanisms lead to impact noise and squeal noise, which are dedicated to specific studies. The other mechanisms are the aerodynamic noise—mainly due to the vehicle speed—and the machinery noise—generated by powered machines such as the electric or diesel motors, the transmission, the cooling fan and so on. Each of these sources of noise can affect people in the vicinity of railway networks.

Several and excellent books and chapters of book treat these problems in a general way, with focus on high-speed network (e.g. [5]). However, very few analyses have been done in the case of urban areas. In the authors' opinion, this created a kind of paradox that it will be explained in the next sections.

This chapter aims at reviewing recent investigations related to railway noise and vibration in urban areas and is divided into three parts:

- In a first section, the problem of railway-induced ground vibration is presented as along with two approaches—experimental and theoretical—to assess the ground vibration level, with particular focus on the modelling approach. The importance of a detailed vehicle model emphasized and the limitations of some prediction tools for urban areas are presented.
- In a second section, the problem of reducing railway noise is used to illustrate the classical approach to noise control. The various origins of airborne noise are presented, including rolling noise, aerodynamic noise and curve squeal noise, with emphasis to the main contributors in urban areas.

http://dx.doi.org/10.5772/66571

- In a last section, the structural noise coming for the building and ground vibration is presented. This is motivated by the misunderstanding of the people, confusing noise and vibration effects. The comfort is evaluated, followed by its effect on building performance.

2. Ground vibration and structural assessment

2.1. Position of the problem

The generation of vibrations is a consequence of the vehicle forces passing from the rotating wheels into the track. These forces depend on the moving vehicle's weight (static contribution, often called quasi-static effect) and surface irregularities at wheel and rail surfaces (representing the dynamic contribution). They contribute to the propagation of vibrations outwards from the track. The vibration level experienced is a function of this force, depending on the amplification factors of each track and soil component (all other locations within the track, soil or nearby structures), as a function of the excitation frequency. Therefore, it is imperative that both effects are well evaluated in the ground vibration assessment.

Much of the research into railway-induced ground vibrations has focused on the effect of high-speed trains on the environment. This was motivated by the so-called supercritical phenomenon which appears when a train travels close to the soil Rayleigh wave speed (critical speed depending on the soil flexibility, which may be close to that of conventional high-speed lines). Despite the large vibration levels generated by these lines which are underlain by soft soils, the distance d between the track and neighbouring structures is relatively large and the vibration attenuates rapidly. In the case of railway traffic, the attenuation is associated with a power law of the form d^{-q}, where q lies between 0.5 and 1.1, depending on the soil configuration [6]. The situation is significantly different in the case of urban transit, due to the presence of local defects which induce elevated localized vibrations (dynamic effect). In the past few years, some studies have emerged that are focused on the vehicle effects (*RIVAS* project, with several work packages dedicated to some mitigation measures for the vehicles [7], *CarboVibes* project focusing on freight railway lines [8]). However, by quantifying all the research projects in railway-induced ground vibration, there is a distinct lack of studies on analysing the effect of local defects on ground vibration. Despite this lack of attention, many ground-borne vibration complaints in urban environments are due to local rail and wheel surface defects (e.g. switches, rail joints, etc.).

As suggested in [9], an analogy between railway-induced ground vibrations and vibration isolation concepts can be established. When a force $f(t)$ is applied on a mechanical system, a part of this is transmitted to the foundation, depending on the characteristics of the isolation (**Figure 1(a)**). On the other hand, when a motion $u(t)$ undergoes the foundation, the equipment has also a motion $x(t)$ depending on the equipment isolator system (**Figure 1(b)**). In the railway, the vehicle/track/soil interaction is associated with the first case, as the force is defined by the wheel/rail interaction, with the quasi-static and dynamic contributions. The role of isolator is played by the track, which has the role of dispatching the forces through the discrete supports (sleepers).

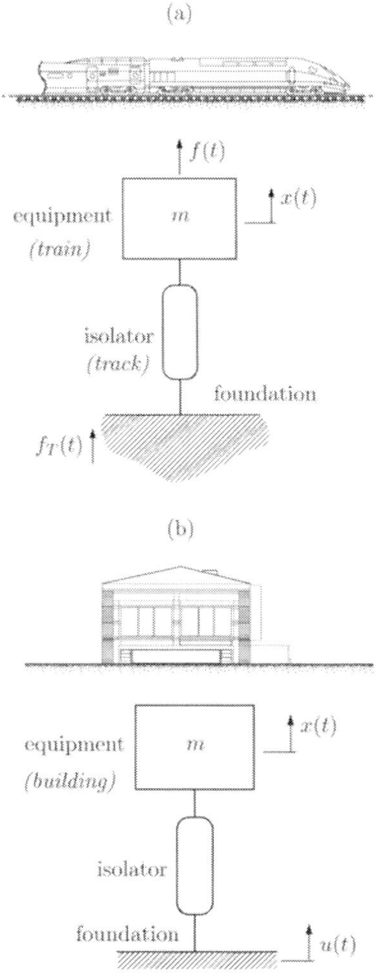

Figure 1. Schematic diagrams of the vibration isolation systems: an analogy with the railway-induced ground vibration [9]. (a) Vibration isolation where force $f(t)$ is applied by the equipment and force $f_T(t)$ is transmitted to the foundation (vehicle/track/soil dynamics). (b) Vibration isolation where motion $x(t)$ is imposed at the foundation and motion $x(t)$ is transmitted to the foundation (soil/structure interaction).

Physical experiments were the conventional means that researchers used to evaluate the effects of vehicles on their surrounding neighbourhoods. However, several cost and physical limitations remain: time and budget constraints, the difficulty involved with investigating a single effect and in cases where the site to be tested does not yet exist. Despite this, the acquisition of experimental data is interesting because it can be used to establish empirical models and validate existing or in-development prediction models. It also serves to illustrate the essential physical interpretations gathered from experience on real lines over the last 20 years. Although many experiments are freely available in the case of high-speed trains, the case of urban railway presents few available and complete studies. However, measurement remains a quick approach for vibration evaluation when the site for analysis exists. If the site is not yet created, preliminary studies and impact surveys can be used but they are limited to other sites of similar composition.

2.2. Nature of urban networks

The vibration generated by the railway therefore depends on the type of vehicle (or network) and the quality of the rolling surface. **Figure 2** illustrates three train types according to their network. It is important to reiterate that this level classification depends not only on the vehicle speed but also on the network type. The various train/track models were classified according to their main excitation mechanisms. High-speed trains generate ground vibrations that are mainly dependent on quasi-static track deflection (effect of a moving constant axle loading), because the high-speed lines are typically characterized by very high-quality-rolling surfaces. This hypothesis is, however, valid when the vehicle speed is lower than a theoretical critical track/soil velocity (often close or greater than 500 km/h). On the contrary, a low speed and a relatively high density of singular rail surface defects (such as rail joints, crossings or switching gears) characterize the light-transit vehicles (LRT) (e.g. trams or metros). As pointed out in [10], the dynamic track deflection (induced by the dynamic interaction between the train and the track) is a main contributor to ground-wave generation. Between these two extreme cases, a combination of contributors experienced on both high-speed and urban railway lines concerns the domestic intercity trains travelling at moderate speeds. A non-negligible influence on ground vibrations is associated to quasi-static track deflection, in addition to the effects due to local defects.

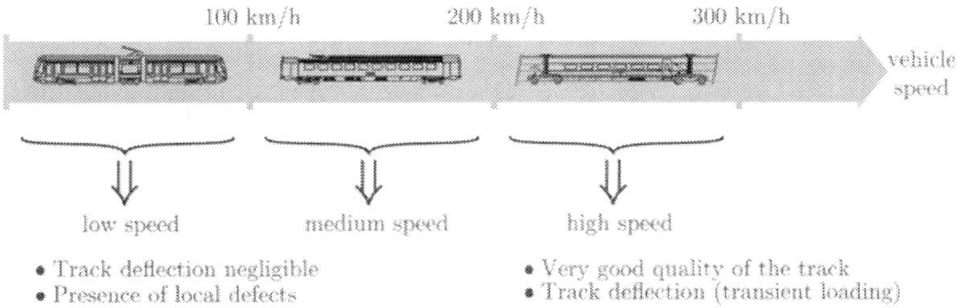

Figure 2. Main contribution to dynamic vehicle/track and soil interactions [10].

Regarding the soil modelling, various methodologies currently exist: (semi-) analytical approach, finite element method (FEM), boundary element method (BEM), among others. The FEM and BEM can be modelled as 2D or 3D problems, depending on the assumed hypothesis. When the boundary conditions mimicking the soil infinity are well defined, the FEM represents an interesting approach due to its ability to describe the soil geometry (layer, tunnel, etc.) in detail and to easily include other structures (e.g., buildings). Compared to coupled FEM-BEM (e.g. [11]), this offers a single approach to model the soil. Furthermore, FEM software packages are already widely used in engineering.

To understand the generation and the propagation of vibrations generated by trains, we illustrate two cases in **Figure 3**: the first one assumes distributed irregularities along the track alignment; the second case is devoted to the presence of a local defect (such as rail joint, switches, crossover, turnout, etc.).

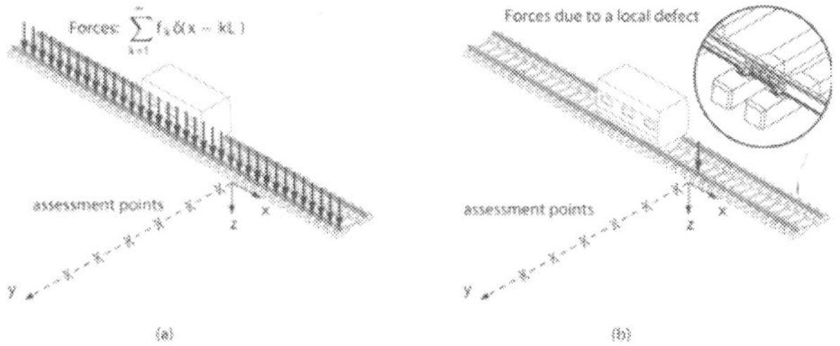

Figure 3. Generation of train forces: (a) for distributed source and (b) for local source of excitation.

In the case of distributed irregularities along the track alignment, the forces issued from the interaction between each wheel set j and the rail can be considered as the sleepers reaction covering a large distance and the excitation can be defined as the summation of the effects of each force f_k acting through the k-th sleeper in the neighbourhood

$$f_{\text{exc},j} = \sum_{k=1}^{\infty} f_k \delta(x - kL) \tag{1}$$

where L is the sleeper bay and δ the Dirac delta function. The resulting vibrations at several distances (assessment points) from the track result from the summation of the effects of each force (often called line source vibration).

In the case of a local defect (**Figure 4**), the ground vibration near railway lines is the result of the interaction of the railway vehicle and the track when the train is running over a local defect in the rail. Therefore, it is relatively reasonable to consider the single force acting on the wheel/rail-defect contact point as the only contributor to railway vibration

$$f_{\text{exc},j} = f_{\text{wheel/rail}} \tag{2}$$

Notice that the force defined by Eq. (2)—acting at the wheel/rail interface—has a location different from the force defined by Eq. (1)—at the track/soil interface, to be compliant with the physical phenomena. Notice also that wheel flats and more generally any defect on wheel-rolling surface, are particular defects since the effect is reproduced every wheel rotation. The periodic effect of wheel-flat impact affects the whole track.

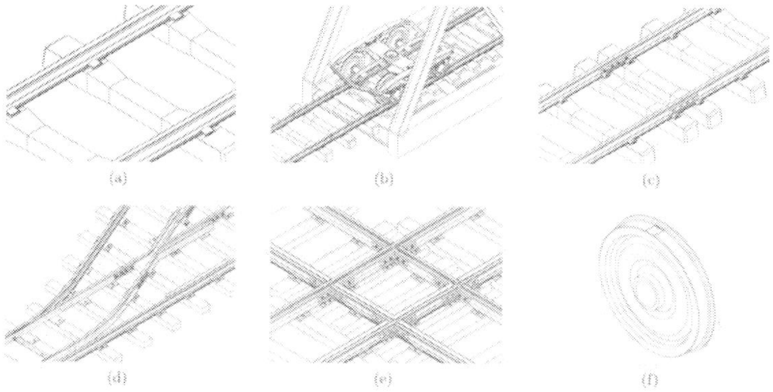

Figure 4. Overview of possible surface defects encountered in practice: (a) reference (no defect), (b) foundation transition, (c) rail joints, (d) turnout system, (e) crossing location and (f) wheel flat.

2.3. Ground-wave problem: a brief history of vibration prediction methods

The aim of a comprehensive ground-borne vibration model is to determine the required mitigation measures in order to guarantee, under examination, along these extensions, that the allowable ground-borne vibration levels in nearby buildings are met. Prediction models abound the literature. For specific situations (transition zones), dedicated approaches are available for estimating the track dynamics. As shown in **Figure 5**, the first prediction models used a simple point source load to simulate the effect of a moving train on a track and to understand the high level of vibrations associated to the supercritical phenomenon. This was first used by Krylov [12]. Following this, many researchers have exclusively focused on high-speed lines, neglecting other cases at lower speeds. Naturally, these models were adapted to be more accurate, including the effect of track unevenness ('random axle loads', e.g. [13]). With the intent to further research the vehicle and track interaction exerted by the wheel and rail irregularities, complete vehicle/track/soil models were proposed, by defining the vehicle with lumped masses (and presented, incorrectly, as multibody models) connected by spring and damper elements representing the suspension system. The effect of detailed vehicle models was clearly discussed in [10]. The aim of this modelling approach was to be more reliable; however, the conclusion was that an accurate description can instead be obtained in simulation by considering only the unsprung and semi-sprung (bogies) masses of the train [14]. In parallel, Kouroussis et al. [15] demonstrated the benefits of including a complete model in the simulation of the ground vibration propagation induced by railway vehicles, that is, the frequency content of ground vibrations involves the signature of the dynamic modes of the vehicle and includes the effect of the sprung mass (car body). Both results are, at a first estimate, contradictory, but the studied cases were different: in [14], the vehicle speed was relatively high (218 km/h), whereas in [15], the train studied was a tram travelling at low speed (30 km/h) running over a local defect.

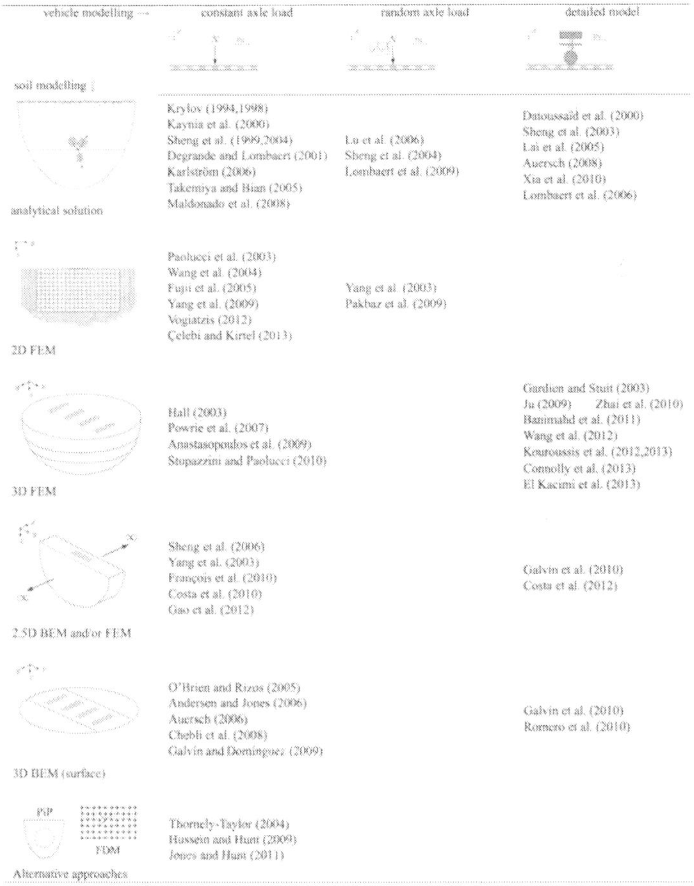

Figure 5. Classification of recent railway-induced ground vibration models (the complete list of references can be found in Ref. [10]).

Regarding the soil modelling, analytical approaches proved their efficiency to simple case. Due to computational burden, numerical approaches were preferred these last. More particularly, BEM offers an attractive way to model infinite medium such as the soil. However, its quasi-exclusive use in the frequency domain limits the possibility to include nonlinearities and, since it uses Green's functions to efficiently calculate vibration propagation at large offsets, only simple geometries/configurations can be assessed. The FEM is an alternative method that has

http://dx.doi.org/10.5772/66571

gained wide acceptance in structural and vibration modelling. It has been used widely for railway vibration problems due to its versatility and to the possibility to model complex geometries. With the increase of computational capabilities these last years, FEM became on the same ranking than BEM. Moreover, the possibility to explicitly model structures/buildings close to the line makes it accessible to urban area problems.

Generally speaking, all the types of model offer valuable information. As pointed out by International Standard Organization (ISO) 14837-1 standard [16], the general circumstances of interest generally define the type of model (**Figure 6**). In the early stage of design, preliminary engineering models offer a rapid way to quantify the order of magnitude of the vibrations felt in the neighbour of railway lines. For advanced design, the need of a detailed model is obvious. Although the ISO 14837-1 standard does not provide any recommendation about which method to use in all the railway cases, it gives useful information about the frequency range within the assessment needs to be made: between 1 and 500 Hz for the effect on the buildings, from 1 to 80 Hz for the evaluation of human exposure to whole-body vibration and up to 200 Hz for sensitive equipment and sensitive tasks.

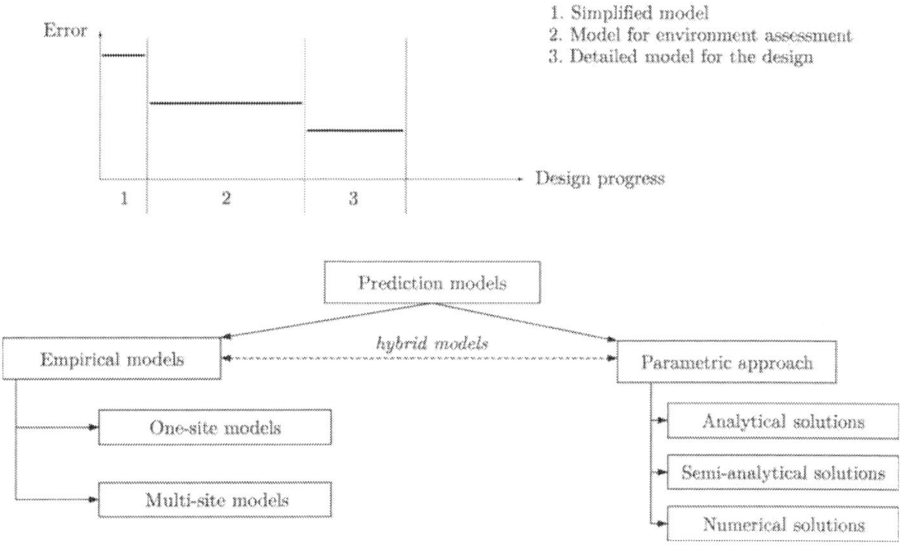

Figure 6. Synopsis of the types of model and the acceptable error in the design process.

2.4. The case of urban traffic: the 'railway paradox'

Several studies have been undertaken in order to evaluate potential vibration-mitigation measures (e.g. trenches [17]) and their effect on urban environments. The research on high-speed trains was motivated by the aforementioned supercritical phenomenon. Despite the large vibration levels generated by these lines, which are underlain by soft soils, the distance between the track and its neighbouring structures is relatively high and the vibration amplitude attenuates rapidly. The situation is significantly different in the case of urban transit for numerous reasons such as:

- the distance *d* between track and buildings is relatively close;
- the contribution of the vehicle's weight and speed (quasistatic effects) is generally low and
- the presence of local defects induces elevated vibrations (dynamic effects) with a different power law.

These differences have produced some contradictory works (that the authors may call 'railway paradox'). The train constructors perform in-depth analysis of the vehicle dynamics by quantifying the vehicle's stability, comfort, behaviour on curved tracks, 3D wheel/rail interactions and motion with complex nonlinear suspensions. This is generally undertaken using multibody simulation (MBS) software tools (using commercial packages such as ADAMS, SIMPACK or Madymo) working with detailed models. The simulation of a complete process that takes into account the track and the soil is not performed: the track and therefore the soil, is modelled to be rigid. Recently, the possibility for these packages to couple the vehicle MBS model with an FEM model of the track using either co-simulation techniques (e.g. [18]) or modal reduction [19] was investigated. Currently, the track/soil vibrations are rarely considered from the initial design stages, even though it is the ideal moment to make ground vibration assessments and to analyse potential vibration-mitigation solutions. On the other hand, train/network operators consider only axle loads from the vehicle. The main reason for this discrepancy is certainly the different approaches adopted by these methodologies: MBS for the vehicle (almost always calculated in time domain) and FEM/BEM for the track/soil subsystem (static analysis or steady-state dynamics, usually calculated in the frequency domain).

This attitude produced a way of thinking to approach the issue based on the soil, considering it to be the principal cause of high-ground vibration levels (e.g. in high-speed lines, it is not the train which has an excessive speed, but the soil which has a low rigidity). Therefore, vibration-reduction measures on the transmission path (track-soil-receiver) have received considerable attention in recent years [20]. Although they represent a sustainable noise and vibration-mitigation measure (ideal candidates for retrofitting existing lines as their installation does not require track closures), they do not include the possibility to act directly on the problem source (in mechanics or in acoustics, it is well known that the first way to solve a problem is to act directly on its source). To illustrate this, a recent example that Kouroussis et al. studied [21] focused on the effect of localized railway defects in urban areas. Although wave number domain-modelling approaches are well suited to predict vibration levels on standard railway lines due to track periodicity, the time domain approach was preferred for non-periodic and localized defects. A fairly accurate description of the interaction between the track and the vehicle was modelled. The main contributions were to model the wheel/rail contact using a nonlinear contact algorithm and to use a detailed 2D vehicle model in the presence of wheel/rail discontinuities. In [22], the potential vibration effect of a flat spot located on a single wheel of a tram is demonstrated. By changing the vehicle and the studied speed range [21], very small levels of vibration are observed for the wheel flat. It was also shown that the type of defect has a significant influence on the levels of vibration. Clearly, the difference between the studied vehicles was revealed and provided a clear requirement for further work on more comprehensive models of the vehicle. Another studied example [23] showed the drawback of using models that were limited to the prediction of only the vertical wheel/rail forces and their interaction with the track and the surrounding ground. The effect of horizontal vibrations due to the presence of rail joints was numerically underestimated, indicating that the dynamic behaviour in the longitudinal direction of the track should also be considered.

2.5. Source of vibration

As aforementioned, the source of vibration arises from the contact between the wheels and the rail. Any imperfection in the rolling surface creates a dynamic effect amplified by the vehicle dynamics and the track/soil response. The presence of local defects induces elevated localized vibrations. In urban area, these local defects are the main contribution of ground vibration because the quasi-static effect is often negligible due to the low speed of the vehicle. The key challenge is to limit the impact of these defects by redesigning the surface shape (e.g. rail joints need to be smoothed by adapting or creating transition zones to avoid abrupt changes in the rolling surface).

Regarding the railway ground vibration models, a linear contact law is often assumed for the vehicle/track coupling by considering small variation around the nomination penetration between the wheel and the rail. This hypothesis is available when the surface imperfection is non-existent or very small. However, when important variations in the contact point are present, the complete Hertz's theory is necessary to accurately predict the interaction forces at wheel/rail contact points [21].

2.6. Evaluation and vibration control

Human perception and building damages due to vibrations are the two main issues needed to be analysed in typical vibration studies [20]. Inhabitant health and comfort may be affected by vibrations that can also affect the structural integrity of buildings due to imposing important dynamic loads. Therefore, engineers need to evaluate all possible damages ensuring that the level of vibrations in building does not cause negative effects on people's comfort. Several vibration standards and recommendations exist, with the most important ones presenting hereafter. The most important ones are the following:

- As the main reference for comfort evaluation, the international standards ISO [24] are often retained. A root-mean-squared (*rms*) value $\breve{a}_W$ is calculated and describes the smoothed vibration amplitude by supposing that the human body responds to an average vibration amplitude during a recorded time $0 \leq t \leq T$

$$\breve{a}_w = \sqrt{\frac{1}{T}\int_0^T a_w^2\,(t)dt} \tag{3}$$

where a_w is the weighted acceleration derived from the time history of the acceleration at the studied location. Guidelines are given for the effect of vibrations on comfort and perception with valuable limits defining grades of various magnitudes of reaction to vibrations. The associated standard [24] less describes the effects on health by giving only two bounds (probable risk if above the upper limit, improbable risk below the lower limit). No additional information is provided for the case when the calculated value lies within the intermediate region.

- The recommendations [25] of the United States Department of Transportation (USDT) on the assessment of potential vibration impacts resulting from high-speed train lines use a decibel scale

$$V_{dB} = 20\log_{10}\frac{v_{\text{rms}}}{5.10^{-8}} \tag{4}$$

where v_{rms} is the root-mean-square amplitude of the velocity time history. Both comfort and structural damages are estimated with this single indicator.

- The German standards DIN4150-2 [26] is used in Germany, in Belgium and some other European countries. A weighted time-averaged signal is defined by

$$KB_F(t) = \sqrt{\frac{1}{\tau}\int_0^T KB^2(\xi)\,e^{-\frac{t-\xi}{\tau}}\,d\xi} \tag{5}$$

where the weighted velocity signal $KB(t)$ is obtained by flowing the original velocity signal through a high-pass filter. The integration time τ to run the averaging is equal to 0.125 s, which allows taking into account transient phenomena such as impacts or shocks that would otherwise be masked if a simple *rms* operation was performed.

- The Swiss and German standards SN640 312a [27] and DIN4150-3 [28] are based on the peak particle velocity *PPV*, defined as the maximum of absolute velocity, to assess the building damages.

All these baselines represent the most used assessment guidelines for measurement and interpretational methodologies.

3. Airborne and ground-borne noise in the vicinity of urban rail networks

The field of railway noise is too large to be presented in a detailed way. Nowadays, noise in urban area is usually managed by national and regional stakeholders. In Europe, the Environmental Noise Directive (END) 2002/49/EC requires European Union Member States to

http://dx.doi.org/10.5772/66571

determine the exposure to environmental noise through strategic noise mapping and to elaborate action plans in order to reduce noise pollution, where necessary. The END [29] has aims to

- define a common homogeneous approach in order to prevent, reduce or avoid the harmful effects due to exposure on environmental noise, including annoyance, on a prioritized basis and
- to provide a basis for developing mitigation measures on all major sources, with emphasis on road, rail and aircraft vehicles and infrastructure, industrial equipment and mobile machinery.

Regarding rail noise in particular, specialized dose-response relationships are needed for new sources of noise such as high-speed railways, or metropolitan underground and superficial tram and metro networks, in order to quantify the impact of additional factors such as technical characteristics of the source, its operation mode with emphasis to speed, the implementation of quiet facades, the influence of nearby green areas, the number and distribution of high-level noise events and spectral aspects (e.g. low-frequency noise). There is need to establish valid dose-response curves for cardiovascular response during sleep and noise taking into account the source characteristics and especially vehicle speed. In urban conditions, railway vehicle passes with a relatively low speed, so traction and rolling noise mainly affect the sound pressure level (SPL) (**Figure 7**).

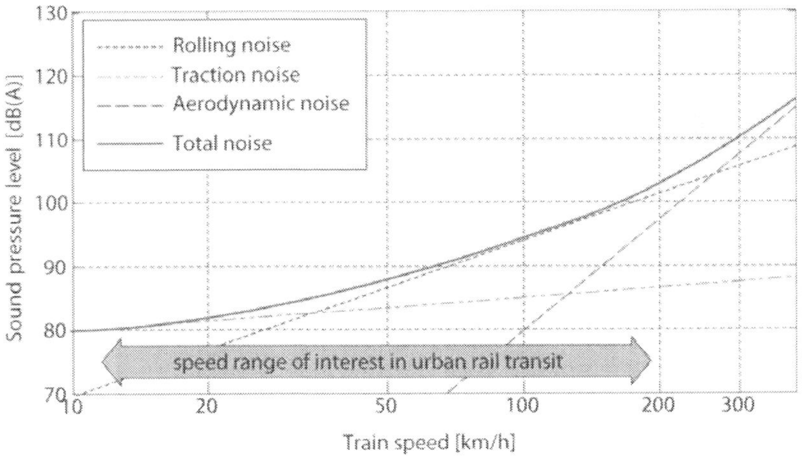

Figure 7. Speed relation for the three noise sources: sound pressure level as a function of train speed.

According to the Directive 2002/49/EC (annex II), the assessment methods and noise indicators for environmental train noise referred in Article 6 [29] are presented hereafter. The day-evening-night level L_{den} in decibels (dB) is defined by the following formula:

$$L_{\text{den}} = 10\log_{10}\frac{12.10^{\frac{L_{\text{day}}}{10}} + 4.10^{\frac{L_{\text{evening}}+5}{10}} + 8.10^{\frac{L_{\text{night}}+10}{10}}}{24} \tag{6}$$

introducing

- the A-weighted long-term average sound level L_{day} determined over all the day periods of a year,
- the A-weighted long-term average sound level L_{evening} determined over all the evening periods of a year and
- the A-weighted long-term average sound level L_{night} determined over all the night periods of a year,

as defined in the international standards ISO 1996-2:1987 [30]. The night-time period is 8 h for the noise indicator L_{night}. A year is a relevant year as regards the emission of sound and an average year as regards the meteorological circumstances and an assessment point is the same as for L_{den}. The L_{den} assessment point is located on a specific height according to the recommendations. In the case of strategic noise mapping, the assessment points must be at 4.0 ± 0.2 m (3.8–4.2 m) above the ground and at the most exposed façade. In such case, the external wall facing onto and nearest to the specific noise source is retained although other choices may be made for other purposes. According to this directive for Member States that have no national computation methods or Member States that wish to change the computation method, the Netherlands national computation method ('Reken-en Meetvoorschrift Railverkeerslawaai 96', or RMR) is recommended. This method provides two different calculation schemes, SRM I (simplified scheme) and SRM II (detailed scheme). The conditions under which each of the schemes can be used are finely described by the method, in order to determine which method to use for the purpose of strategic noise mapping following the Directive 2002/49/EC [31].

In practice, the obligations which Directive 2002/49/EC imposes on the Member States and the European Commission address noise from road, rail and air traffic and industrial installations in agglomerations. Legislation on sources is complementary to the END as reducing the contribution to noise at source obviously reduces the exposure at the receiver. Recently, the European Commission in cooperation with the European Union Member States developed a common framework for noise assessment methods, called CNOSSOS-EU, which represents a harmonized and coherent approach to assess noise levels from the main sources of transport-induced noise (road traffic, railway traffic, aircraft and industrial). In 2015, CNOSSOS-EU became a new European Union Commission Directive (based on a revised Annex II of the END) and will be mandatory for all European Union Member States after 31 December 2018 [31]. A focus will be paid on a number of implementation challenges that should be faced in the context of current and potential European Union environmental noise policy developments in view of CNOSSOS-EU becoming fully operational in the European Union Member States.

The aforementioned Dutch railway noise computation method RMR has its own emission model; however, the emission model remains as the original. Prior to the calculation of an 'equivalent continuous sound pressure level' generated by all vehicles that use a specified section of railway line and follow the appropriate service, guidelines should be either placed in the 10 railway vehicles categories provided in the Dutch emission database (**Table 1**).

Category	Train description
1	Block-braked passenger trains
2	Disc-braked and block-braked passenger trains
3	Disc-braked passenger trains
4	Block-braked freight trains
5	Block-braked diesel trains
6	Diesel trains with disc brakes
7	Disc-braked urban subway and rapid tram trains
8	Disc-braked InterCity and slow trains
9	Disc-braked and block-braked high-speed trains
10	Provisionally reserved for high-speed trains of the ICE-3 (M) type

Table 1. Railway vehicle categories provided by the Dutch emission RMR database.

According to SRM I, emission values in dB(A) are determined as follows:

$$E = 10 \log\left(\sum_{c} y 10^{E_{nr,c}/10} + \sum_{c} y 10^{E_{r,c}/10}\right) \tag{7}$$

where $E_{nr,c}$ is the emission term per rail vehicle category for nonbraking trains, $E_{r,c}$ the emission term for braking trains, c the train category and y the total number of categories present. The emission values per rail vehicle category are determined from

$$E_{nr,c} = a_c + b_c \log v_c + 10 \log Q_c + C_{b,c} \tag{8}$$

$$E_{r,c} = a_{r,c} + b_{r,c} \log v_c + 10 \log Q_{br,c} + C_{b,c} \tag{9}$$

where the standard emission values a_c, b_c, $a_{r,c}$ and $b_{r,c}$ are provided in RMR. Q_c and $Q_{br,c}$ are the mean number of non-braking and braking units of the railway vehicle category concerned, respectively, v_c is the mean speed of passing railway vehicles (making the distinction between braking and non-braking units) and $C_{b,c}$ is a correction factor. SRM II suggests that the emission values per octave band are determined for each train category and for different sound source heights (up to five heights). The emission of the specified section of railway line is calculated taking into account the passage of different train categories with the emission factor in octave band i to be calculated as follows:

$$L_{E,i}^{h} = 10 \log\left(\sum_{c} n 10^{E_{nb,i,c}^{h}/10} + \sum_{c} n 10^{E_{br,i,c}^{h}/10} \right) \quad (10)$$

where

$$E_{br,i,c}^{h} = a_{br,i,c}^{h} + b_{br,i,c}^{h} \log v_c + 10 \log Q_{br,c} + C_{bb,i,m,c} \quad (11)$$

$$E_{nb,i,c}^{h} = a_{i,c}^{h} + b_{i,c}^{h} \log v_c + 10 \log Q_{br,c} + C_{bb,i,m,c} \quad (12)$$

Additional parameters are introduced: $a_{br,i,c}^{h}$, $b_{br,i,c}^{h}$, $a_{i,c}^{h}$ and $b_{i,c}^{h}$ are emission terms for train category in braking and nonbraking conditions, for octave band i at height h. $C_{bb,i,m,c}$ is also a correction factor, including the presence of the track disconnection, the track discontinuity and the rail roughness.

The new common framework for noise assessment methods (CNOSSOS-EU) recently developed by the European Commission in co-operation with the EU Member States is to be applied for strategic noise mapping, represents a harmonized and coherent approach to address and assess noise levels from the main sources of noise—including railway traffic—based on state-of-the-art knowledge and resulted from an intensive collaboration, exchange of data and evaluation procedure via a formal process at both policy and scientific/technical levels [32]. In the new environmental noise calculation method, a vehicle is defined as any single railway sub-unit of a train moving independently. All sub-units are grouped into a single vehicle. The existing tracks may also differ due to different acoustic properties. The overall track properties are defined by two acoustically essential parameters, for example, the railhead roughness and the track decay rate, according to ISO 3095:2013 [33], as well as the radius of curvature of the track.

The different equivalent rail airborne noise line sources are placed at different heights and at the centre of the track. The equivalent sources include various categories of physical sources as follows:

- the rolling noise (including rail- and track-base vibration wheel vibration as well as superstructure noise of the freight vehicles),

- the traction noise,
- the aerodynamic noise,
- the impact noise (from crossings, switches and junctions),
- the squeal noise and
- the noise due to bridges and viaducts.

In the new methodology [31, 32], the sound power emission assessment for railway traffic noise is analogue to road traffic noise. The noise sound power emission of a specific track type to fulfil a series of requirements is described in the vehicle and track classification, in terms of a set of sound power per each vehicle $L_{w,0}$. Furthermore, the noise emission of a traffic flow on each track is represented by a set of two source lines with relative directional sound power per metre and per frequency band. This corresponds to the sum of the sound emissions due to the individual vehicles passing by taking into account the time spent by the vehicles in the railway section (for stationary vehicles). The directional sound power per metre and per frequency band, due to all the vehicles passing by each track section on the track type (j), is defined as follows:

- for each frequency band (i);
- for each given source height (h) (for sources at (1) h = 0.5 m and (2) h = 4.0 m) and is the energy sum of all contributions from all vehicles running on the specific j-th track from:
    - the vehicle types (t),
    - the speeds (s),
    - the running conditions (constant speed) (c) and
    - the source types (as presented above, e.g., rolling, impact, squeal, traction, aerodynamic and bridge noise) (p).

To calculate the directional sound power per metre (input to the propagation part) due to the average mix of traffic on the jth track section, the following formula is used:

$$L_{W',\mathrm{eq},T,\mathrm{dir},i} = 10\log\left(\sum_{x=1}^{X} 10^{L_{W',\mathrm{eq},\mathrm{line},x}/10}\right) \tag{13}$$

where T is the reference time period for which the average traffic is considered and X is the total number of existing combinations of i, t, s, c, p for each j-th track section. $L_{W',\mathrm{eq},\mathrm{line},x}$ is the xth directional sound power per metre for a source line of one combination of t, s, c, p on each jth track section. It takes into account the index t for vehicle types on the jth track section, the index s for train speed, the index c for running conditions: 1 (for constant speed) or 2 (idling), the index p for physical source types: 1 (for rolling and impact noise), 2 (curve squeal), 3 (traction noise), 4 (aerodynamic noise) or 5 (additional effects).

If a steady flow of vehicles Q/hour is assumed, with an average speed v, there will be an equivalent number of Q/v vehicles per unit length of the examined railway section, on average, at each moment in time. The noise emission of the vehicle flow in terms of directional sound power per metre $L_{W',\text{eq, line}}$ (expressed in dB/m (ref. 10^{-12} W)) is integrated by

$$L_{W',\text{eq,line},i}(\psi,\phi) = L_{W,0,\text{dir},i}(\psi,\phi) + 10\log\left(\frac{Q}{1000\ v}\right)(\text{for } c = 1) \tag{14}$$

where Q is the average number of vehicles per hour on the j-th track section and $L_{W,0,\text{dir},i}$ the directional sound power level of the specific noise (rolling, impact, squeal, braking, traction, aerodynamic, other effects) of a single vehicle in the angular directions (ψ, ϕ) defined with respect to the vehicle's direction of movement.

The vehicle contribution and the track contribution to rolling noise are separated into four essential elements: wheel roughness, rail roughness, vehicle transfer function to the wheels and to the superstructure (vessels) and track transfer function. Not including one of these four parameters would prevent the decoupling of the classification of tracks and trains [38].

4. Examples

Three examples are retained in this chapter to illustrate the complexity of ground vibration in urban areas.

4.1. T2000 circulating in Brussels: the effect of vehicle design

The first case is related to Belgian urban public transport company who replaced the old PCC7000 trams by the new T2000 LRV in Brussels Capital Region (Belgium). The T2000 LRV tram was developed by Bombardier Transport and is defined as a multicar tramway characterized by a low-floor design. This imposes that bogies involve independent rotating wheels and the motors are mounted directly inside the wheels. This example is based on experimental studies, after having observed important vibratory nuisances in the neighbourhood of this new tram. Among all the studies initiated by national projects to alleviate the vibratory level in the surrounding buildings, the research work [34] focused the effect of the roughness or local unevenness such as a rail defect on the soil vibration level.

Experiments have been performed by measuring the vibrations induced by the passing on an artificial local defect. Simulations completed the study showing the surprising results that a decrease vehicle speed from 20 to 30 km/h reveals a reduction of the vibratory level. The developed model allowed verifying this trend by simulating other cases considering a larger velocity range. **Figure 8** illustrates this statement and plots the vertical *PPV* as a function of the distance from the track and the tram speed. It turns out that the *PPV* regularly decreases not only with the distance but also with the speed.

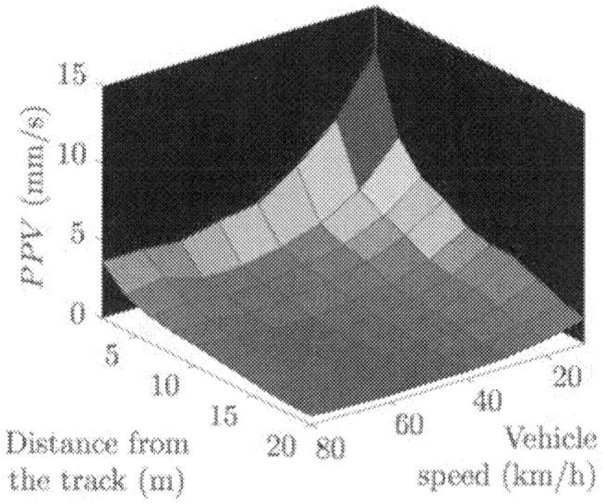

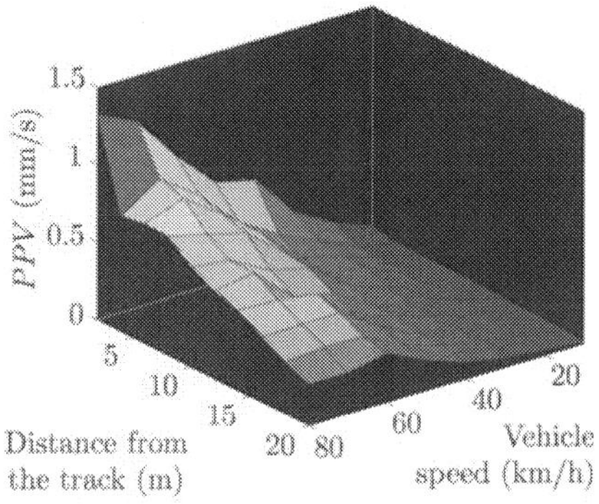

Figure 8. Vertical peak particle velocity PPV calculated from T2000 LRV passing as a function of the distance from the track and of the tram speed [34]. (a) During the passing on the local defect and (b) during the passing on a rough rail (without local defect).

4.2. Vibration and ground-borne noise generated by the passing of vehicles over turnouts —the case of Athens Metro

The second example illustrates the challenge in old and dense urban centres as the municipality of Athens (Greece) and the associated ground vibration problems of underground networks with respect to the protection of the cultural heritage in the case of archaeological area and museum. Intensive measurement campaigns were performed over the last decade, combined with predictive calculation in order to evaluate the efficiency of floating slabs as mitigation measures. Light-rapid-transit underground and surface networks in urban conditions represent a substantial reduction of air pollutants emissions. Indeed, this sustainable means of transportation offers a way to decrease the number of cars and heavy vehicles (i.e. buses) circulating in a road network. However, as aforementioned, an increased level of vibration transmitted to buildings in close proximity is often observed. This example illustrates this important adverse effect of Athens metro operation. It was observed that ground-borne noise and vibration in buildings was the result of the dynamic impact forces generated at the wheel-rail interface coupled to wheel and rail irregularities. A direct transmission of ground-borne vibration induced by metro traffic was clearly identified: ground-borne vibrations excite the foundation walls of nearby buildings, beneath the ground.

Figure 9 shows the results regarding the evaluation of anti-vibration performance of the floating slabs at several crossover locations [35, 36]. Additional measurements were recorded in order to verify that the vibration level at the closest buildings to the crossover locations was also below the fixed limit. To quantify the gain brought by the mitigation measure, the insertion loss factor IL is usually used. The ratio of the vibration level between the unisolated and the isolated track is defined:

$$IL = 20\log_{10}\frac{v_{unisolated}}{v_{isolated}} \tag{15}$$

where $v_{\text{unisolated}}$ and v_{isolated} are the corresponding vibration velocity amplitudes. **Figure 9** shows in which frequencies the response decreased due to the insertion of a floating. This example demonstrates how mitigation measures can be tuned to efficiently reduce the generated ground vibration levels in a specified frequency range.

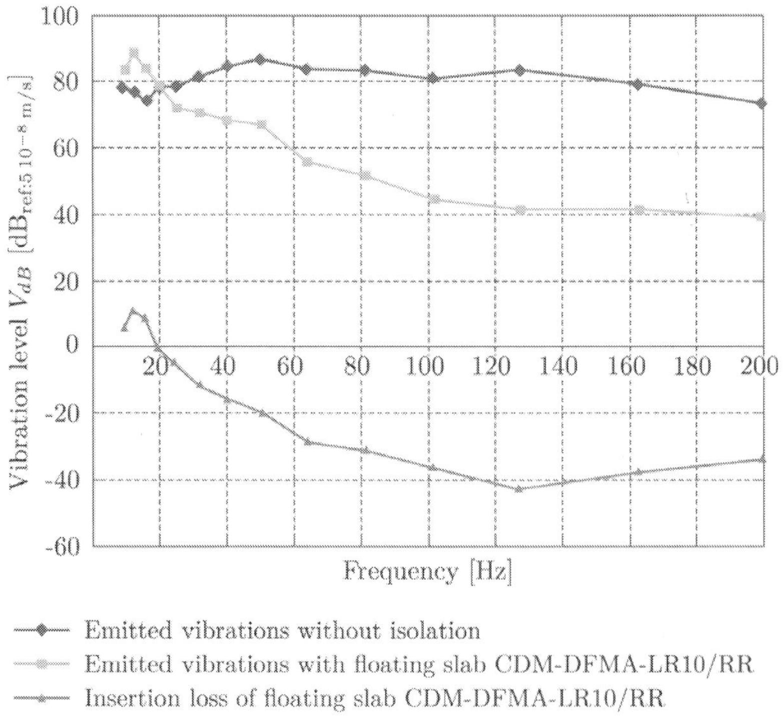

Figure 9. Comparison of 1/3 frequency analysis of vibration velocity with and without floating slabs at Athens metro extensions crossover locations [35].

4.3. Airborne noise in the vicinity of urban rail networks—the Quiet-Track Project in Athens Metro line 1

The overall objective of the Quiet-Track project [37] is to provide efficient solutions by including track-based noise-mitigation systems and maintenance schemes. The project focused on the development and validation of performance solutions for reduction of track-related noise and on providing track noise management tools. Existing solutions were combined to yield an overall attenuation of at least 6 dB(A). This was done by simulation: the simulation of the combined effect of these solutions was completed by the implementation and the validation of the results in the network of Attiko Metro line 1, in Athens. The track was composed by twin-block concrete sleepers rigidly embedded in a concrete slab track. An existing outside concrete slab—ballastless system—track revealed high airborne noise and a combination of existing solutions was evaluated for noise reduction. Three distinct actions regarding possible mitigation measures were investigated, both on individual and on combined bases [38]:

- *Action 1*: Absorbing panels placed on the track, close to the source, between the rails aiming to mitigate contributions of both the wheels and the rails by influencing sound waves that are normally reflected.
- *Action 2*: Noise-reflective barrier was considered next to the track along the protective fence, preventing sound propagation by directly reflecting the sound waves, achieving an important level of noise reduction.
- *Action 3*: Rail dampers were also considered. Their application was confined to the track itself, influencing only the contribution of the rail. This action is to be considered only in cases where the rail contribution dominates or it is equally important as the wheel contribution.

Numerical simulations were also implemented combining different software tools for noise prediction [39]. The procedure was validated in the network of Attiko Metro line 1 where the selected noise-mitigation measures were installed. The expected attenuation of sound pressure level *SPL* (before and after mitigation measures) using the insertion loss factor *IL* in dB described by

$$IL = SPL_{after} - SPL_{before}, \tag{16}$$

was therefore determined and afterwards compared to the measured noise reduction. A full program of acoustic 'initial situation' noise measurements, of train normal conditions operation, was ALSO executed in order to establish the acoustic performance of the installed RHEDA system before any noise-mitigation measure implementation. The International Standards ISO 3095:2013 [33] specifies the conditions for obtaining reproducible and comparable measurement results of levels and spectra of noise for vehicles operating on rails or other types of fixed track. It was implemented using a class 1 multi-channel noise analyser [39]. **Figure 10** shows the sound pressure level in dB(A) computed by the software IMMI. The overall noise level before installation was 78.1 dB(A), meanwhile after installation of all three mitigation measures was reduced to 68.9 dB(A), resulting in an overall gain of 9.2 dB(A). The measured sound pressure levels for all cases are also presented, suggesting an overall good correlation between both calculated/simulated and measured values. **Table 2** summarizes the results for all cases.

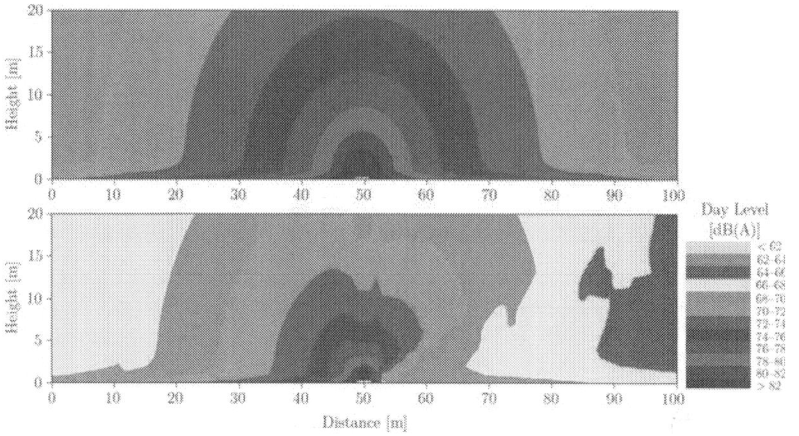

Figure 10. Sound pressure level SPL in dB(A) before (reference) and after all mitigation measures were in place [39].

The Quiet-Track Project in Athens Metro NOISE-MITIGATION MEASURES ACTIONS	Overall SPL [dB(A)]	
	Measured	**Simulated**
Reference	**78.1**	**78.1**
Absorbing panels	**76.1**	**76.0**
Absorbing panels + noise barrier	**69.5**	**70.7**
Absorbing panels + noise barrier + rail dampers	**68.0**	**68.9**

Table 2. Comparison of measured and simulated overall sound pressure levels [39].

5. Concluding remarks

As suggested in [1], as vibration prediction models become more complex, they achieve higher accuracy at the expense of an increase in software computational requirements. Four cornerstones of a vibration prediction model are usually interlinked, as shown in **Figure 11**. For instance, a simple empirical model presents a negligible execution time with a high usability. Extensive training is thus not required and the parameter requirements are low. This results in few input parameters which require investigation prior to execution. However, these advantages balance with a poor accuracy of such simplified model.

The most challenging aspect in the study is to develop models for urban cases. The nature of wheel/rail contact must be analysed in-depth in order to develop a comprehensive vehicle/track model. This offers a way to treat complex problems encountered in practice where the train interacts with important local defects and to quantify the effect of these defects according to their size and shape for any possible situation.

Compared to vibration assessment, noise evaluation benefits to a certain degree of maturity. CNOSSOS-EU framework will offer an even better and efficient way to evaluate the noise level within strategic noise mapping and to propose adequate environmental noise-mitigation actions. Especially regarding airborne noise-mitigation comparing solutions such as absorbing panels on the track, noise barriers next to the track and rail dampers, considerable positive results may be achieved, ensuring also a very good correlation between predicted and measured noise levels. The developed technology can, therefore, be used without restrictions by all concerned such as engineering companies working in the field, consultants, contractors, operators and infrastructure managers and cities. Therefore, a dissemination strategy for a wide-spread information transfer was also set up, to ensure that the above results need to be integrated into existing standards, informing the relevant stakeholders who will implement the results, maximizing therefore the market uptake by external dissemination activities towards other nonparticipating bodies to monitor and implement specific national needs of the light-rapid-transit operators.

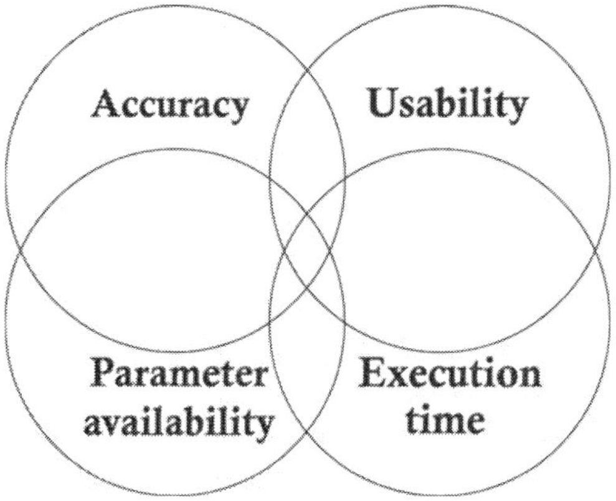

Figure 11. The four desirable characteristics of a vibration prediction model [1].

Author details

Konstantinos Vogiatzis[1*] and Georges Kouroussis[2]

*Address all correspondence to: kvogiatz@uth.gr

1 Laboratory of Transportation Environmental Acoustics, School of the Civil Engineering, University of Thessaly, Pedion Areos, Volos, Greece

2 Department of Theoretical Mechanics, Dynamics and Vibrations, Faculty of Engineering, University of Mons, Place du parc B-7000, Mons, Belgium

References

[1] Connolly, D. P.; Kouroussis, G.; Laghrouche, O.; Ho, C. & Forde, M. C. Benchmarking railway vibrations—track, vehicle, ground and building effects. Construction and Building Materials. 2015;92:64–81. DOI: 10.1016/j.conbuildmat.2014.07.042

[2] Vogiatzis, K. Protection of the cultural heritage from underground metro vibration and ground-borne noise in Athens Centre: The case of the Kerameikos Archaeological Museum and Gazi Cultural Centre. International Journal of Acoustics and Vibration. 2012;17(2):59–72. DOI: 10.20855/ijav.2012.17.2301

[3] Kouroussis, G.; Pauwels, N.; Brux, P.; Conti, C. & Verlinden, O. A numerical analysis of the influence of tram characteristics and rail profile on railway traffic ground-borne noise and vibration in the Brussels Region. Science of the Total Environment. 2014;482–483:452–460. DOI: 10.1016/j.scitotenv.2013.05.083

[4] Stiebel, D.; Muller, R.; Bongini, E.; Ekbald, A.; Coquel, G.; Alguacil, A.A. Definition of reference cases typical for hot-spots in Europe with existing vibration problems. Rivas Project SCP0-GA-2010-265754 (deliverable D1.5) ed. Report to the EC; 2012.

[5] Thompson, D. Railway noise and vibration: mechanisms, modelling and means. Oxford, UK: Elsevier Ltd.; 2009.

[6] Auersch, L. & Said, S. Attenuation of ground vibrations due to different technical sources. Earthquake Engineering and Engineering Vibration. 2010;9(3):337–344. DOI: 10.1007/s11803-010-0018-0

[7] Nielsen, J.; Mirza, A.; Cervello, S.; Huber, P.; Müller, R.; Nelain, B. & Ruest, P. Reducing train-induced ground-borne vibration by vehicle design and maintenance. International Journal of Rail Transportation. 2015;3(1):17–39. DOI: 10.1080/23248378.2014.994260

[8] Waddington, D.; Woodcock, J.; Smith, M.G.; Janssen, S. & Waye, K.P. CargoVibes: human response to vibration due to freight rail traffic. International Journal of Rail Transportation. 2015;3(4):233–248. DOI: 10.1080/23248378.2015.1076623

[9] Kouroussis, G.; Van Parys, L.; Conti, C. & Verlinden, O. Prediction of ground vibrations induced by urban railway traffic: an analysis of the coupling assumptions between vehicle, track, soil and building. International Journal of Acoustics and Vibration. 2013;18(4):163–172. DOI: 10.20855/ijav.2013.18.4330

[10] Kouroussis, G.; Connolly, D.P. & Verlinden, O. Railway induced ground vibrations - a review of vehicle effects. International Journal of Rail Transportation. 2014;2(2):69–110. DOI: 10.1080/23248378.2014.897791

[11] Galvín, P.; François, S.; Schevenels, M.; Bongini, E.; Degrande, G. & Lombaert, G. A 2.5D coupled FE-BE model for the prediction of railway induced vibrations. Soil Dynamics

and Earthquake Engineering. 2010;30(12):1500–1512. DOI: 10.1016/j.soildyn.2010.07.001

[12] Krylov, V. V. Effect of track properties on ground vibrations generated by high-speed trains. Acustica-acta Acustica. 1998;84(1):78–90.

[13] Lu, F.; Gao, Q.; Lin, J. H. & Williams, F. W. Non-stationary random ground vibration due to loads moving along a railway track. Journal of Sound and Vibration. 2006;298(1–2):30--42. DOI: 10.1016/j.jsv.2006.04.041

[14] Costa, P. A.; Calçada, R. & Cardoso, A. S. Influence of train dynamic modelling strategy on the prediction of track-ground vibrations induced by railway traffic. Journal of Rail and Rapid Transit. 2012;226:434–450. DOI: 10.1177/0954409711433 686

[15] Kouroussis, G.; Verlinden, O. & Conti, C. Efficiency of resilient wheels on the alleviation of railway ground vibrations. Journal of Rail and Rapid Transit. 2012;226(4):381–396. DOI: 10.1177/0954409711429210

[16] International Organization for Standardization. ISO 14837-1: Mechanical vibration—Ground-borne noise and vibration arising from rail systems—Part 1: General guidance. 2005.

[17] Connolly, D.; Giannopoulos, A.; Fan, W.; Woodward, P. & Forde, M. Optimising low acoustic impedance back-fill material wave barrier dimensions to shield structures from ground borne high speed rail vibrations. Construction and Building Materials. 2013;44:557–564. DOI: 10.1016/j.conbuildmat.2013.03.034

[18] Dietz, S.; Hippmann, G. & Schupp, G. Interaction of vehicles and flexible tracks by co-simulation of multibody vehicle systems and finite element track models. Vehicle System Dynamics. 2002;37(Supplement 1):372–384. DOI: 10.1080/00423114.2002.11666 247

[19] Kassa, E. & Nielsen, J. C. Dynamic train-turnout interaction in an extended frequency range using a detailed model of track dynamics. Journal of Sound and Vibration. 2009;320(4-5):893–914. DOI: 10.1016/j.jsv.2008.08.028

[20] Connolly, D. P.; Marecki, G.; Kouroussis, G.; Thalassinakis, I. & Woodward, P. K. The growth of railway ground vibration problems—a review. Science of the Total Environment. 2016;568:1276–1282. DOI: 10.1016/j.scitotenv.2015.09.101

[21] Kouroussis, G.; Connolly, D. P.; Alexandrou, G. & Vogiatzis, K. Railway ground vibrations induced by wheel and rail singular defects. Vehicle System Dynamics. 2015;53(10):1500–1519. DOI: 10.1080/00423114.2015.1062116

[22] Alexandrou, G.; Kouroussis, G. & Verlinden, O. A comprehensive prediction model for vehicle/track/soil dynamic response due to wheel flats. Journal of Rail and Rapid Transit. 2016;230(4):1088–1104. DOI: 10.1177/0954409715576015

[23] Kouroussis, G.; Florentin, J. & Verlinden, O. Ground vibrations induced by InterCity/InterRegion trains: A numerical prediction based on the multibody/finite element

modeling approach. Journal of Vibration and Control. 2016;22(20): 4192–4210. DOI: 10.1177/1077546315573914.

[24] International Organization for Standardization. ISO 2631-2: Mechanical vibration and shock—Evaluation of human exposure to whole-body vibration—Part 2: Vibration in buildings (1 Hz to 80 Hz). 2003.

[25] U. S. Department of Transportation (Federal Railroad Administration). High-speed ground transportation. Noise and vibration impact assessment, Technical Report 293630-1. Office of Railroad Development Washington; 1998.

[26] Deutsches Institut für Normung. DIN 4150-2: Structural vibrations—Part 2: Human exposure to vibration in buildings. 1999.

[27] Schweizerische Normen-Vereinigung. SN-640312a: Les ébranlements—Effet des ébranlements sur les constructions. 1992.

[28] Deutsches Institut für Normung. DIN 4150-3: Structural vibrations—Part 3: Effects of vibration on structures. 1999.

[29] European Parliament. Directive 2002/49/EC of the Council of 25 June 2002 relating to the assessment and management of environmental noise. OJ L 189 ed. 2002.

[30] International Organization for Standardization. ISO 1996-2: Acoustics—Description and measurement of environmental noise—Part 2: Acquisition of data pertinent to land use. 1987.

[31] European Commission. Directive 2015/996 of 19 May 2015 establishing common noise assessment methods according to Directive 2002/49/EC of the European Parliament and of the Council. OJ of the European Communities ed. 2015.

[32] Kephalopoulos, S. & Paviotti, M. Common noise assessment methods for Europe (CNOSSOS-EU): implementation challenges in the context of EU noise policy developments and future perspectives. In: K. Vogiatzis, G. Kouroussis, M. Crocker, M.Pawelczyk, editors. Proceedings of the 23rd International Congress on Sound and Vibration (ICSV23); 10–14 July; Athens (Greece). 2016.

[33] International Organization for Standardization. ISO 3095: Acoustics—Railway applications—Measurement of noise emitted by railbound vehicles. 2013.

[34] Kouroussis, G.; Verlinden, O. & Conti, C. Contribution of vehicle/track dynamics to the ground vibrations induced by the Brussels tramway. In: P. Sas, B. Bergen, editors. ISMA2010 International Conference on Noise and Vibration Engineering; September 20–22; Leuven. 2010. pp. 3489–3502.

[35] Vogiatzis, K. & Kouroussis, G. Prediction and efficient control of vibration mitigation using floating slabs: Practical application at Athens metro lines 2 and 3. International Journal of Rail Transportation. 2015;3(4):215–232. DOI: 10.1080/23248378.2015.1076622

[36] Carels, P., Ophalffens, K. & Vogiatzis, K. Noise and vibration evaluation of a floating slab in direct fixation turnouts in "Haidari & Anthoupoli extentions" of Athens Metro lines 2 and 3. Ingegneria Ferroviaria. 2012;67(6):533–553.

[37] Vogiatzis, K. & Desanghere, G. Deliverable 4.1: Procedure to select the combination of solutions for optimal noise performance, WP4: Development and validation of high performance solutions for reduction of track related noise—Task 4.1. QUIET-TRACK Quiet Tracks for Sustainable Railway Infrastructures (FP7-SST-2013-RTD-1); 2014. DOI: http://www.quiet-track.eu

[38] Vogiatzis, K. & Desanghere, G. Deliverable 4.2: Validation section in network of Attiko Metro with optimal combination of existing solutions, WP4: Development and validation of high performance solutions for reduction of track related noise—Task 4.1. QUIET-TRACK Quiet Tracks for Sustainable Railway Infrastructures (FP7-SST-2013-RTD-1); 2014. DOI: http://www.quiet-track.eu

[39] Vogiatzis, K. & Vanhonacker, P. Noise reduction in urban LRT networks by combining track based solutions. Science of the Total Environment. 2016;568:1344–1354. DOI: 10.1016/j.scitotenv.2015.05.060

Chapter 5

Characterisation of Airborne Particulate Matter in Different European Subway Systems

Vânia Martins, María Cruz Minguillón,
Teresa Moreno, Luís Mendes,
Konstantinos Eleftheriadis, Célia A. Alves,
Eladio de Miguel and Xavier Querol

Additional information is available at the end of the chapter

http://dx.doi.org/10.5772/65364

Abstract

Air quality sampling campaigns in three European subway systems (Barcelona, Athens and Oporto) were conducted in order to characterise particulate matter (PM) to better understand the main factors controlling it. PM mass concentrations varied among the European subway platforms, and also within the same underground system, this being mainly associated to differences in the design of the stations and tunnels, system age, train frequency, ventilation and air-conditioning systems, commuter's density, rails geometry and outdoor air quality. PM concentrations displayed clear diurnal patterns, depending largely on the operation and frequency of the trains and the ventilation system. Chemically, subway $PM_{2.5}$ on the platforms consisted of iron, carbonaceous material, crustal matter, secondary inorganic compounds, insoluble sulphate, halite and trace elements. Fe was the most abundant element, accounting for 19–46% of the bulk $PM_{2.5}$, which is generated mainly from mechanical wear at rail-wheel-brake interfaces. A source apportionment analysis allowed the identification of outdoor (sea salt, fuel-oil combustion and secondary aerosol) and subway sources on platforms. The use of air-conditioning inside the trains was an effective approach to reduce exposure concentrations, being more efficient removing coarser particles. PM concentrations inside the trains were greatly affected by the surrounding (i.e. platforms and tunnels) air quality conditions.

Keywords: metro, platforms, trains, subway aerosol, indoor air quality, exposure, commuting

1. Introduction

Urban air quality plans incentivise the use of public transport to abate atmospheric emissions from road vehicles. In this context, underground subway systems with electric trains are especially desirable as they are energetically efficient and contribute to relieve surface traffic congestion; hence, it is considered one of the cleanest public transport systems. The subway system is one of the major transport modes in most metropolitan areas worldwide, due to its convenience, safety and high speed. Its high capacity in terms of the number of daily commuters makes it an environmentally friendly alternative.

Particulate matter (PM) in the underground subway microenvironments is of great concern since many people spend considerable time commuting on a daily basis, and the exposure to this pollutant in the subway systems has been linked to adverse health effects [1–3]. Studies have indicated, with few exceptions, that PM concentrations are usually higher in these underground environments than in outdoor ambient air, as these environments are a confined space poorly ventilated promoting the concentration of PM entering from the outside atmosphere in addition to that generated internally in the system [[4] and references therein]. Particles in the subway system are mainly generated by the motion of trains, movement of commuters and subway staff, air ventilation, and works of maintenance and construction. Most particles in this environment are produced at the rail-wheel-brake interfaces by friction and mechanical wear processes, and at the interface between the current collectors attached to trains and the power-conductive materials providing electricity. Additional PM sources are provided by the erosion of construction materials and their subsequent resuspension [5–9]. Subway trains are typically powered either by an overhead catenary, involving the electrical current being drawn through the contact material of the pantograph, or by a third rail in which the current passes to the train via a contact shoe. Both coarse and fine particles are produced during shearing between wheels, rails and brakes, and ultrafine particles can be generated during the high temperatures resulting from friction at interfaces between these components, in some cases leading to vaporisation of the materials [7, 10].

Perhaps, more interesting than the bulk mass concentration of PM is the fact that these particles have peculiar physico-chemical characteristics specific to the subway environment, being loaded with ferruginous particles commonly accompanied by other elements such as Mn, Si, Cr, Cu, Ba, Ca, Zn, Ni and K [4–6, 11–16]. The considerable amount of Fe in the subway stations is mainly generated from mechanical friction and wear processes between rails, wheels and brakes [5, 15, 17, 18]. Wear and friction processes initially produce iron-metal particles that react with oxygen in the air resulting in the formation of iron oxides [5, 15, 19].

In any case, the concentration and chemical composition of subway particles depend on various factors, such as outdoor air quality; differences in the depth and design of the stations and tunnels; system age; composition of wheels, rails, brakes and power supply materials; braking mechanisms; power system; train speed and frequency; passenger densities; ventilation and air-conditioning systems; cleaning frequency and other operational conditions [23]: [16, 17, 20–23]. Knowing the chemical composition of PM in a subway platform is an essential prerequisite for understanding the indoor air quality of the subway system and subsequently to assess remediation measures. Moreover, the chemical composition of PM derived by sample analysis can be further utilised for risk-assessment studies and although components such as the trace metals represent typically only about 1% of the total PM, they can play a critical role in the source identification [24, 25].

The aim of this study was to characterise personal exposure to PM while commuting, including the waiting time on the platform and travelling inside the trains, in the subway systems of three European cities, to better understand the main factors controlling air quality in this environment. The work was based on air quality campaigns following the same sampling, measurement and analytical methods, and data treatment.

2. Studied subway systems

Three European subway systems were selected: Barcelona (Spain), Athens (Greece) and Oporto (Portugal), although with main focus on Barcelona.

The Barcelona subway system is an extensive network of electrified railway lines that runs mostly underground. The network has 8 lines (numbered L1–L5 and L9–L11), 139 stations, 102.6 km of track (January 2016), carries around 376 million passengers each year, and is managed by *Transports Metropolitans de Barcelona* (TMB) [8, 15]. The platforms and tunnels are equipped with mechanical forced ventilation which favours the air exchange between the indoor and outdoor environments. All trains are operated using a rigid overhead catenary for power supply and run from 5:00 h until midnight every day, with additional services on Friday nights (finishing at 2:00 h of Saturday) and Saturday nights (running all night long), with a frequency between 2 and 15 min, depending on the day (weekend or weekday), subway line and time of day. Trains from all lines are equipped with an efficient air-conditioning system that works continuously throughout the year to maintain a comfortable temperature, but with higher intensity during the warmer period.

The Athens subway system is run by *Urban Rail Transport S.A.* and is used for the transportation of nearly 494 million passengers per year in the city of Athens. Line 1 was a conventional steam railway constructed in 1869, which was converted to electrical railway in 1904, and runs almost entirely above ground. Lines 2 and 3 opened in 2000 and are mostly underground (a portion of the L3 is a suburban rail line that runs above ground). The total length of the network is 82.7 km and includes 61 stations (January 2016). Trains run from around 5:30 until 00:30 h, with a frequency of 4–5 min during the rush hours and 7–15 min in the off-hours. The trains are provided with air-conditioning system and there is the ability to open the windows. The network uses electric trains equipped by both contact shoes and pantographs, which in the underground sections runs on third rail and in the above ground uses overhead catenaries.

The Oporto subway system is part of the public transport system of Oporto. *Metro do Porto, S.A.*, is engaged in the operation and maintenance of the subway system. The network has 6 lines (LA, LB, LC, LD, LE and LF) with the first line being opened in 2002. Currently, the system has an extension of 67 km with a total of 81 operational stations, 14 of which are underground

(January 2016). The system is underground in central Oporto (8 km of the network) and above ground into the city's suburbs, carrying about 57 million passengers per year. Trains run every day from 6:00 until 1:00 h with a frequency from 5 to 19 min, and are equipped with air-conditioning system. The power supply system is a solid overhead catenary line.

3. Experimental methods

3.1. Platform measurements

3.1.1. Intensive campaigns

In the case of the Barcelona subway study, four stations with distinct designs belonging to different lines were selected for the intensive campaigns: Joanic (L4), Santa Coloma (L1), Tetuan (L2) and Llefià (L10). The architectural design of the stations and tunnels is different for each station. In both Joanic and Santa Coloma stations, there is one wide tunnel with two central rails served by lateral platforms, although in Joanic the two rails are separated by a middle wall. In the case of Tetuan and Llefià, there is a narrower tunnel with just one platform and one rail, although with the major difference that in the more modern Llefià station the platform is separated from the rail by a glass wall within which is embedded a platform screen door (PSD) system (**Table 1**).

Two 1-month intensive campaigns were carried out at each of the stations during two periods: warmer (2 April–30 July 2013) and colder (28 October 2013–10 March 2014). Ventilation protocols were different in the warmer and the colder periods, which allowed to ascertain seasonal differences (**Table 1**).

$PM_{2.5}$ samples were collected with a high-volume sampler (30 $m^3 h^{-1}$, HVS, Model CAV-A/MSb, MCV) on quartz microfibre filters. Sampling was done daily for 19 h according to the subway working hours (5:00–24:00). Continuous measurements (24 h day^{-1}) with a 5-min time resolution were performed using a light-scattering laser photometer (DustTrak, Model 8533, TSI) for PM_1, $PM_{2.5}$ and PM_{10} mass concentrations and an indoor air quality meter (IAQ-Calc, Model 7545, TSI) for CO_2 and CO concentrations, temperature and relative humidity (RH).

In the case of the Athens and Oporto subway studies, the intensive sampling campaign was carried out at one station in each system, namely Nomismatokopio and Bolhão, respectively. For comparison purposes, the chosen stations had similar platform design: wide tunnel with two rails in the middle, one for each direction, with lateral platforms (**Table 1**). In Athens campaign, $PM_{2.5}$ samples were collected using a high-volume sampler, similar to the one used in Barcelona. In Oporto campaign, a high-volume sampler (TE-5200, Tisch Environmental Inc.) operating at a flow of 67.8 $m^3 h^{-1}$ was used to collect coarse ($PM_{2.5-10}$) and fine ($PM_{2.5}$) particles, although only the $PM_{2.5}$ data were used in this study. The particles were collected daily on quartz microfibre filters during the subway operating hours (from 5:30 to 00:30 h in Athens and from 6:00 to 01:00 h in Oporto). Field-filter blanks were also collected. A DustTrak and an indoor air quality meter were simultaneously operated at a 5-min time resolution during 24 h day^{-1}, as in Barcelona's campaign.

Subway system	Station				No. of additional platforms studied	Measurements inside trains (No. of lines)
	Name (line)	Sampling period	Design	Mean train frequency (trains h^{-1})		
Barcelona	Joanic (L4)	2 Apr–2 May 2013 28 Oct–25 Nov 2013	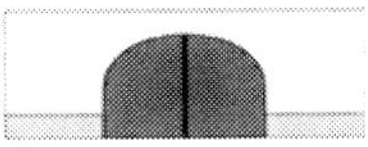	12	24	6
	Santa Coloma (L1)	1 Jul–30 Jul 2013 10 Feb–10 Mar 2014	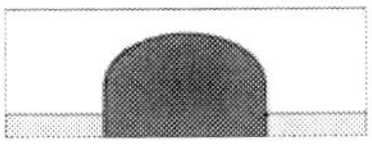	29		
	Tetuan (L2)	2 May–31 May 2013 25 Nov–20 Dec 2013		14		
	Llefià (L10)	31 May–1 Jul 2013 13 Jan–10 Feb 2014	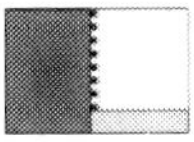	8		
Athens	Nomismatokopio (L3)	28 Apr–19 May 2014	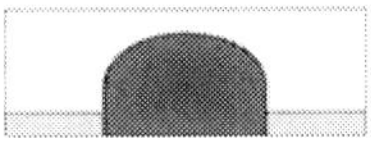	21	5	2
Oporto	Bolhão (LA,LB,LC,LE and LF)	27 Oct–14 Nov 2014	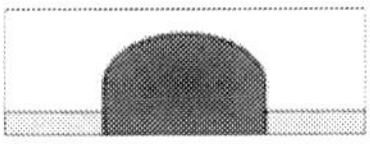	37	5	2

Note:  two-way tunnel railway;  one-way tunnel railway; station platforms; middle wall; glass wall with PSD.

Table 1. Sampling campaigns information.

$PM_{2.5}$ concentrations provided by DustTrak monitor were corrected against the in situ and simultaneous gravimetric $PM_{2.5}$ for each station. Concentrations of PM_1 and PM_{10} were corrected using the same correction factors obtained for $PM_{2.5}$. Therefore, in this study only the $PM_{2.5}$ concentrations are used in absolute terms, whereas the PM_1 and PM_{10} concentrations are only used to assess relative variations.

In the three subway studies, sampling and monitoring devices were placed at a distance from the commuters' access-to-platform point and behind a light fence for safety protection. The exact location chosen on each platform was typically a compromise between the availability of power supply, the need not to obstruct passenger movement and the preference to make the equipment as inconspicuous as possible.

3.1.2. Additional monitoring

Additional platforms were selected to study the temporal and spatial variations in the PM_X concentrations along the platforms. A total of 24 platforms from Barcelona subway system, 5 from Athens subway system, and 5 from Oporto subway system were studied (**Table 1**). Note that these stations include the aforementioned stations selected for the intensive campaigns (four in Barcelona, one in Athens and one in Oporto). Out of the 24 stations in Barcelona, 4 were new stations (line 10) and the remaining were old stations (lines 1–5), with a preference being given to selecting the most common station designs. This station selection included both those with double rail tracks with (4 stations) and without (14 stations) a middle wall, and a single rail track with (4 stations) and without (2 stations) a PSD system [8]. The subway stations chosen in Athens both have wide tunnels and two rail tracks, although one has a central platform and the others have two lateral platforms. In Oporto subway system, all stations are double track with lateral platforms.

Measurements were performed at 4 positions approximately equidistant along each platform, during a total of 1 h divided into periods of 15 min (at each of the 4 positions). Additionally, the sampling in the first point was repeated for 5 min after the 4 positions as a control. Real-time PM_1, $PM_{2.5}$ and PM_{10} mass concentrations were registered using a DustTrak monitor set at 5-s time resolution, enabling us to see the effect of trains and commuter's movements. All measurements were carried out during weekdays after 9:00 h to avoid rush hours. The times of trains entering and departing the station were manually recorded to assess possible correlations with the variability of the registered concentrations. The described procedure was conducted twice at each subway platform in Athens and Oporto, and four times in Barcelona (twice during each seasonal period).

3.2. Train measurements

In addition to measuring air quality on platforms, data were also collected from inside trains. In the case of Barcelona, this involved measurements in six different subway lines (L1–L5 and L10), as compared to two lines in both Athens (L2 and L3) and Oporto (LA and LD) (**Table 1**). In each case, the same sampling protocol was adopted. PM_1, $PM_{2.5}$ and PM_{10} mass concentrations were measured using a DustTrak monitor and CO_2 concentrations were monitored by means of an indoor air quality meter in the middle of the central carriage of the train during a two-way trip along the whole length of the subway line [8]. Both instruments were set at a 5-s time resolution. The instrumentation was transported in a bag with the air-uptake inlet placed at shoulder height when sitting. The measurements were carried out after 10:00 h on weekdays, and they were performed twice at each of the selected lines in Athens and Oporto, while they were performed four times in Barcelona (twice during each seasonal period). During the colder period of the Barcelona campaign, the measurements were carried out along the whole length of the line with and without air-conditioning (not possible during warmer period due to passenger's comfort requirements), so that the effect of it on the air quality could be assessed. During each sampling session, a written record was made of each journey, making observations such as when doors opened and closed or (for Athens) if windows were open, or (for Oporto) if the train was travelling above or below ground.

3.3. Outdoor measurements

For comparison purposes, outdoor ambient $PM_{2.5}$ samples were collected concurrently at an urban station at each city. The Barcelona and Athens outdoor measurements were performed using a HVS in the urban background stations of Palau Reial and Demokritos, respectively. The station of Palau Reial is located in the garden of the IDAEA-CSIC at the North-West of the city (41°23′14″ N, 02°06′56″E, 78 m.a.s.l). The Demokritos station is located in NCSR 'Demokritos' campus (37°99′50″ N, 23°81′60″ E, 270 m.a.s.l), at the North-East corner of the Greater Athens Metropolitan Area. The measurements were carried out for 24 h every third day at Palau Reial station, and 19 h (subway operating hours) every second day at Demokritos station. The Oporto outdoor measurements were conducted in the urban traffic station of Francisco Sá Carneiro-Campanhã (41°09′46.10″ N, 08°35′26.95″ W, 147 m.a.s.l), with two low-volume Tecora samplers (TCR, Model 2.004.01) operating a flow of 2.3 $m^3 h^{-1}$. $PM_{2.5}$ samples were collected by both TCR samplers simultaneously for 19 h (coinciding with the subway operating hours) every second day.

4. Sample treatment and analyses

4.1. Filters treatment and weight

Before sampling, quartz microfibre filters were heated in an oven at 200°C for a minimum of 4 h to eliminate the volatile impurities. The filters were equilibrated for at least 48 h in a conditioned room (20°C and 50% relative humidity) and then weighed before and after sampling by means of a microbalance (Model XP105DR, Mettler Toledo). The gravimetric $PM_{2.5}$ mass concentrations were determined dividing the weight difference between the blank and the sampled filter by the volume of air sampled. Once the gravimetric determination was performed, the filters were cut into several sections for subsequent chemical analyses.

4.2. Chemical analyses

The first section of each filter was acid digested and then analysed by inductively coupled plasma-atomic emission spectroscopy (ICP-AES) and ICP-mass spectroscopy (ICP-MS) to obtain concentrations of major and trace elements, respectively. In addition, the standard reference material NIST 1633b was also analysed in a blank filter to check the accuracy of the analysis. The second section was water leached with de-ionised water to extract the soluble fraction and analysed by ion chromatography for determination of soluble anions (Cl^-, SO_4^{2-} and NO_3^-), and by selective electrode for ammonium (NH_4^+). A third portion of the filters was used to measure total carbon (TC) using a thermal-optical method. A detailed description of the analytical methodology has been reported by Querol et al. (2012) [15].

The final ambient air concentrations were calculated after the subtraction of analytical blank values from the corresponding sample concentrations. Detection limits and uncertainties of the determined species concentrations were calculated from the standard deviations from the blank filters analyses alongside the analytical uncertainties [26].

4.3. Source apportionment

After the complete chemical characterisation of $PM_{2.5,}$ a receptor model was applied in order to determine and quantify the sources of atmospheric $PM_{2.5}$ for the Barcelona subway study. The source apportionment was carried out by means of the positive matrix factorisation (PMF) [27] using the US Environmental Protection Agency (US-EPA) PMF 5.0 software. This multivariate receptor model provides estimates of the chemical composition of PM associated with different sources and the mass contribution attributed to each source.

PMF analyses were performed separately for each subway station from the Barcelona system with datasets including both seasonal periods. The species uncertainties were calculated according to Escrig et al. (2009) [26]. For the analysis, all chemical species analysed were summed as the total variable, not taking into account the non-determined mass due to humidity and heteroatoms. Species included in the model were selected based on their signal-to-noise ratio, the percentage of samples above detection limit and their significance (considering their possible presence in this environment).

5. PM mass concentrations

5.1. On platforms

From the extensive characterisation of 24 stations with distinct designs of the Barcelona subway system, a substantial variation in $PM_{2.5}$ concentrations among the stations was observed (hourly averages ranging from 13 to 154 $\mu g\ m^{-3}$) [28]. This variation might be related to the differences in the design of the stations and tunnels, variations in the train frequency, passenger densities and ventilation systems, among other factors, as discussed below. Large variations were also observed in the Athens (22–158 $\mu g\ m^{-3}$; five stations) and Oporto (65–265 $\mu g\ m^{-3}$; five stations) subway systems [8].

In the Barcelona study, the stations composed by a single tunnel with one rail separated from the platform by a wall with PSD (new stations) showed on average lower $PM_{2.5}$ concentrations (around 50%) in comparison with the old conventional stations, which is related to a combination of factors such as (i) the PSD preventing the air from the tunnel entering the platform, (ii) the more advanced ventilation setup and (iii) the lower train frequency [28]. Within the conventional system, the stations with single narrow tunnel and one rail showed on average $PM_{2.5}$ concentrations higher than in stations with one wide tunnel and two rails separated by a middle wall, most probably due to the less efficient dispersion of air pollution, enhancing the accumulation of PM. In the stations with one wide tunnel and two rails without a middle wall, $PM_{2.5}$ concentrations were much more variable. Similarly, Jung et al. [5] reported that at narrow stations there is a larger dependence on strong ventilation to maintain relatively low PM concentrations. Regarding Athens subway system, the $PM_{2.5}$ concentrations in a central station were higher than in a peripheral station (out of the central area of the city), even when both stations belong to the same line (L2), which is probably attributable not only to the age (new station opened in 2013) and location of the station but also to the train frequency (some trains do not run the entire line) and lower number of passengers. Furthermore, measurements in a transfer station (lines 2 and 3 intersect) showed that the $PM_{2.5}$ concentrations were higher in the station platform of L2 than that of L3, probably related to the age of the lines [8].

To compare the three subway systems among them, three stations with similar platform design were selected to minimise other factors influencing the variation of $PM_{2.5}$ concentrations: Santa Coloma in Barcelona, Nomismatokopio in Athens and Bolhão in Oporto. The lowest mean $PM_{2.5}$ concentration (± standard deviation of daily concentrations) was found in Santa Coloma station (58.3 ± 13.7 μg m^{-3}) while the highest mean $PM_{2.5}$ concentration was recorded in Bolhão station (83.7 ± 45.7 μg m^{-3}) (**Figure 1a**). In the Nomismatokopio station, a mean $PM_{2.5}$ concentration of 68.3 ± 11.3 μg m^{-3} was obtained (**Figure 1a**). This range of results may be associated to different ventilation systems, since the Barcelona subway is equipped with mechanical forced ventilation in all its length, whereas in both Athens and Oporto subways only natural ventilation occurs, with air exchange with the outdoor air happening mainly through blast shafts. The mechanical forced ventilation is a relevant factor to improve the air quality within the subway system, as explained below. Moreover, the majority of the underground sections in the Oporto subway system are composed of curved and/or sloping rails, which may imply higher emissions from the rail-wheel-brake interfaces while trains are stopping on the platform and thus producing increased concentrations on the platforms. Train frequency at the sampling site in the Oporto subway is higher than those of both Barcelona and Athens (**Table 1**), as trains from five different lines (LA, LB, LC, LE and LF) converge on Bolhão station using a common platform, whereas in Barcelona and Athens only trains of one line serve each station and consequently the train frequency at the platform is lower [8]. Furthermore, the daily average $PM_{2.5}$ concentrations were much more variable in the Bolhão station than in the other two stations, due to the variable weather conditions, and consequently the $PM_{2.5}$ concentrations in the outdoor ambient air were considerably variable during the sampling period in Oporto. The PM concentrations in the Bolhão station may be particularly affected by the outdoor conditions, since it is followed by an above ground station which favours the air exchange with the exterior.

In general, the mean $PM_{2.5}$ concentrations on the subway platforms were notably higher (between 1.4 and 6.9 times) than those simultaneously recorded in the outdoor ambient air, indicating the presence of indoor particulate sources in the underground stations (**Figure 1a**).

During weekdays, the $PM_{2.5}$ concentrations on the station platforms were considerably higher (1.2–1.5 times) than those measured during weekends, due to the higher number of both commuters and trains. Similar results have been observed in other subway systems [11, 13, 17, 29]. However, considering the three subway systems this difference between the weekdays and weekends in $PM_{2.5}$ concentrations was more pronounced in Bolhão station (Oporto) and less in Nomismatokopio station (Athens), possibly again favoured by the busy environment of Bolhão station with the passage of trains of five lines.

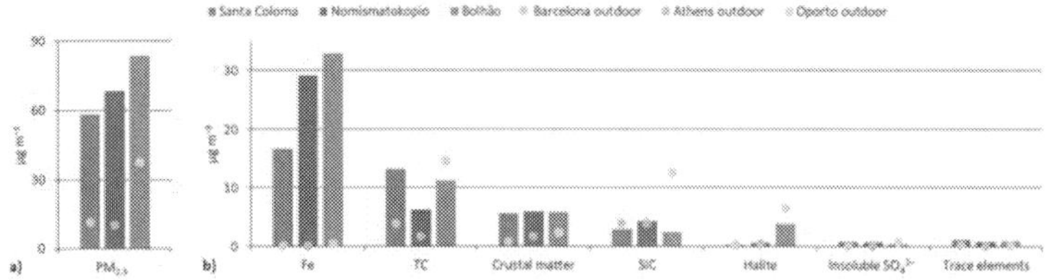

Figure 1. Mean concentrations of $PM_{2.5}$ **(a)** and their elemental components **(b)** in Santa Coloma, Nomismatokopio and Bolhão and in the simultaneous outdoor ambient air (TC, total carbon; SIC, secondary inorganic compounds).

5.1.1. Influence of different ventilation settings

During the extensive campaign in Barcelona where different ventilation protocols were tested, mean $PM_{2.5}$ concentrations on Joanic, Santa Coloma, Tetuan and Llefià subway platforms ranged between 21 and 51 μg m^{-3} in the warmer period, and between 32 and 93 μg m^{-3} in the colder period (**Figure 2**). Seasonal differences among the four stations showed that the concentrations in the colder period were higher and generally more variable than in the warmer period, mainly due to the stronger ventilation in the warmer period that affects the air quality of the subway system, as the weaker ventilation enhances the accumulation of particles in the stations. These results were observed in all the additional platform measurements [28]. Regarding the PM size distribution, the PM_1/PM_{10} and $PM_{2.5}/PM_{10}$ ratios were higher in the warmer period, indicating that the ventilation of the subway system was more efficient removing coarser particles. Thus, PM_1 was the major size fraction composing the PM in the subway system, especially during the warmer period.

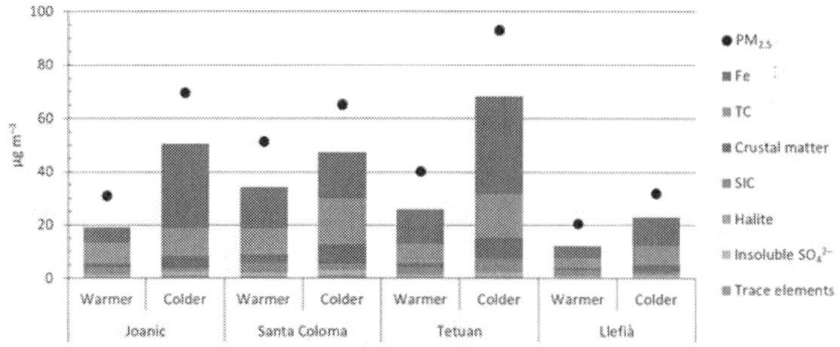

Figure 2. Mean concentrations of $PM_{2.5}$ and the associated elemental components on the Joanic, Santa Coloma, Tetuan and Llefià platforms during the warmer and colder periods (TC, total carbon; SIC, secondary inorganic compounds).

5.1.2. Daily patterns

Similar daily trends were observed among the subway platforms of the three subway systems (**Figure 3**). The $PM_{2.5}$ daily pattern presented a concentration increase in the morning with the arrival of the first trains showing a peak in the morning rush-hour period, which was attributable not only to the influx of commuters (CO_2 generated through exhalation) but also to the higher train frequency; the movement of the commuters leads to the PM resuspension, and the train movement promotes the resuspension of PM and its generation due to the abrasion of rails, wheels, brakes and power supply materials. Afterwards, $PM_{2.5}$ concentration decreased towards a stable concentration until late afternoon. An increase in the $PM_{2.5}$ concentrations was registered again during the evening rush hours, especially in the Bolhão station where the rise in train frequency was very important (increasing approx. 10 trains h^{-1}). In Nomismatokopio, there was no increase in $PM_{2.5}$ concentrations in the evening because both train frequency and influx of commuters decreased during these hours. During the night, there was a continuous decrease in $PM_{2.5}$ concentrations due to transport service interruption for several hours. However, in the Barcelona subway system some outliers during night-time series were generally identified in the conventional stations (Joanic, Santa Coloma and Tetuan), associated with occasional maintenance or cleaning operations (only Santa Coloma is shown as an example in **Figure 3**). The daily patterns evidence that the personal exposure to $PM_{2.5}$ concentrations is dependent on the time of the day used to commute.

In addition to the influence of the train frequency, in the Barcelona subway system, the changes of the ventilation settings along the day had considerable effect in the variations of the $PM_{2.5}$ concentrations on the platforms, particularly in the warmer period, when the ventilation is more intense [28]. It is evident that the impact of train frequency on $PM_{2.5}$ levels only becomes relevant in the absence of strong ventilation. Hence, the daily pattern of $PM_{2.5}$ concentrations in the Barcelona subway system was primarily influenced by the ventilation settings and secondarily by the train frequency [28].

5.1.3. Temporal and spatial variations

Although there were generally day-to-day fluctuations in $PM_{2.5}$ concentrations on the platforms, some temporal and spatial trends were observed along the platforms due to the influence of the ventilation settings but also to the design of the stations and tunnels, location of passengers' access to the platforms, commuter density, as well as to the effect of the passage and frequency of the trains.

The $PM_{2.5}$ concentrations on some platforms varied significantly in short time scales (e.g. an increase of a factor of 3 in less than 30 s), especially in the case of Athens and Barcelona subways [8, 28]. In some cases, the high time resolution measurements evidenced that $PM_{2.5}$ concentrations on the platform increased when the train entered the platform and decreased when it departed. Each train pushes into the station polluted air from the tunnel (by the piston effect) and $PM_{2.5}$ generated by resuspension, and when the train leaves the station the reverse piston effect moves polluted air out of the station, renewing the air on the platform. This effect of passage of trains was especially strong in the new stations (with PSD) and old stations with single rail, although in some stations with two rails without a middle wall this pattern was also observed [28]. The results showed that the PSD in the new stations do not prevent completely PM exchange between the railway and the platform.

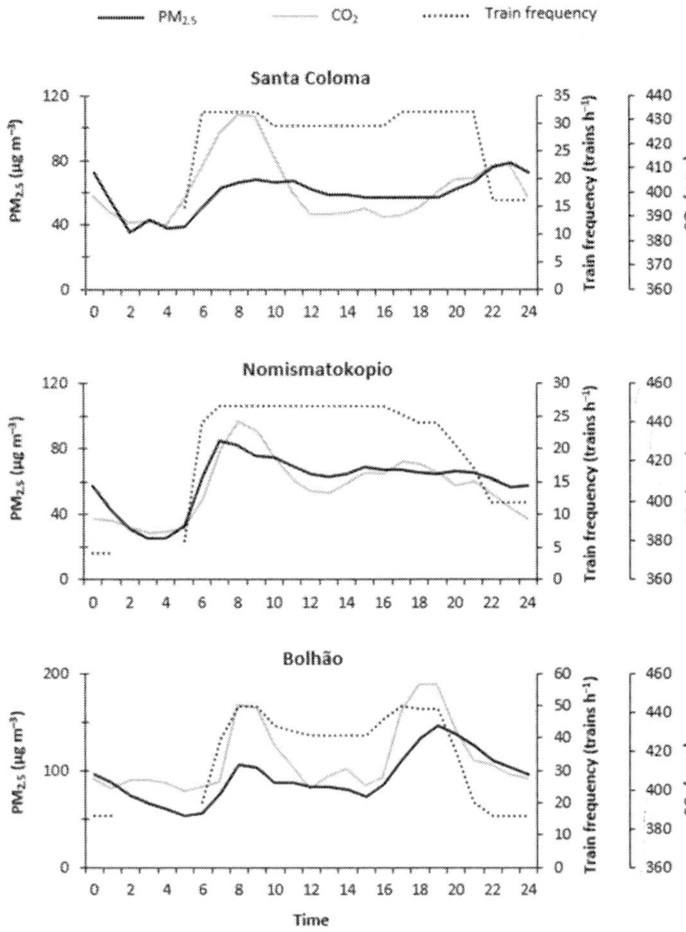

Figure 3. Temporal variation of mean hourly $PM_{2.5}$ and CO_2 concentrations and train frequency in the Santa Coloma, Nomismatokopio and Bolhão subway stations.

In some subway stations in Barcelona, higher $PM_{2.5}$ mass concentrations, especially of coarse particles, were recorded in the train-entry locations and in the areas closer to the commuters' access to the platforms, in comparison with other points on the platform [28]. However, in the Athens and Oporto cases this spatial variation was not clearly observed. Such a variation can be attributed to the turbulence generated by the train's entry, due to the wind blasts produced. In the areas closer to the passengers' access to the platforms, there is also a high probability of $PM_{2.5}$ resuspension, created by the commuters walking and the air flowing in and out of the station.

Moreover, $PM_{2.5}$ concentrations were relatively constant in time and along the platform of some stations. Therefore, in these cases the exposure levels of commuters were very similar when waiting anywhere along the platform.

5.2. Inside trains

The PM_X and CO_2 concentration profiles during trips inside the trains showed dissimilar behaviours. The CO_2 concentrations were clearly driven by the number of passengers inside the train carriages due to exhalation with the maximum influx of people corresponding to stations located in the central area of each city. An increase in the CO_2 concentrations inside the train was sometimes observed when the doors closed and a rapid drop was recorded when the doors opened.

The trains in the three subway systems are equipped with air-conditioning system. The mean $PM_{2.5}$ concentration ranges inside the trains were 19–75, 78–135 and 29–79 μg m^{-3} in Barcelona (six lines), Athens (two lines) and Oporto (two lines), respectively. The study of the use of air-conditioning inside the trains of the Barcelona system evidenced that the air-conditioning provided a clear abatement of PM concentrations, resulting in lower PM_X concentrations (by around 30% for $PM_{2.5}$) and finer particles (PM_1/PM_{10} was around 15% higher), as well as lower variability of PM_X concentrations, than when the air-conditioning was switched off. Additionally, CO_2 exhaled by commuters accumulated inside the trains when air-conditioning was switched off and was less easily removed by the ventilation system compared to PM_X. Generally, the PM_X concentrations along the lines were relatively constant, while short-term peaks were observed after the train doors closed in a number of cases, probably due to turbulence and consequent PM resuspension produced by the movement of passengers inside the trains. In the Athens subway system, carriage windows were usually open, despite the existence of air-conditioning, resulting in an increase in PM_X concentrations inside the trains as they passed through some of the tunnel sections [8]. And thus, the highest PM_X concentrations inside the trains from the three systems were found in the lines belonging to Athens subway system [8]. In the Oporto subway system, the PM_X and CO_2 concentrations inside the trains were generally higher while travelling in the underground than in the above ground sections, where outdoor ambient air entering the trains produced an environmental 'cleaning effect' [8]. Therefore, the PM_X concentrations inside the trains of this subway system are greatly dependent on outdoor ambient air quality.

The PM_X concentrations inside the trains were in general lower than those on the corresponding platforms in the Barcelona and Oporto subway systems, which may be attributed to the air-conditioning system operating inside the trains, and in Oporto also to the predominance of above ground stations along the lines. By contrast, in Athens system, the PM_X concentrations inside the trains were in general higher than those on the platforms since, as stated above, the trains run with most windows open, favouring the entrance of polluted air from tunnels into the trains.

In the Barcelona subway system, the $PM_{2.5}$ concentrations inside the trains in the new line (L10 with PSD) were on average around 50% lower than in the oldest lines (lines 1–5). Thus, the lowest PM_X concentrations were found in the new line both on the platforms and inside the trains, because it is a technologically advanced line with more efficient mechanical ventilation system. Moreover, comparing the real-time measurements performed on the 24 stations with the measurements inside the trains of the six lines, there was the evidence that PM_X levels inside the trains were affected by the surrounding conditions, that is, those on the platforms. **Figure 4** shows that the $PM_{2.5}$ concentrations inside the trains were strongly correlated with the $PM_{2.5}$ concentrations on the corresponding platforms ($R^2 = 0.75$). The lines with high PM concentrations were the first lines in operation and are the busiest ones because they run through the downtown area.

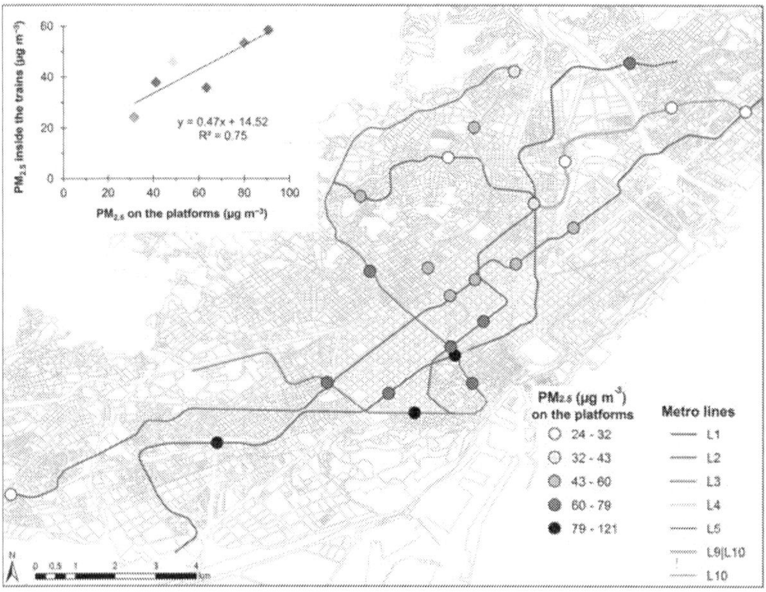

Figure 4. Relation between $PM_{2.5}$ concentrations on the platforms and inside the trains in the Barcelona subway system.

6. Chemical composition of $PM_{2.5}$

The species present in the $PM_{2.5}$ samples can be broadly grouped into seven categories, namely (1) elemental iron (Fe), (2) total carbon (TC), (3) mineral or 'crustal' matter (CM; the sum of Ca, Mg, Al_2O_3, SiO2, CO_3^{2-}, Ti, K and P), (4) secondary inorganic compounds (SIC; the sum of water-soluble sulphate, nitrate and ammonium), (5) halite (NaCl), (6) insoluble sulphate and (7) trace elements. The analysed chemical species accounted for, on average, 59–73% of the total $PM_{2.5}$ on the platforms (**Figures 1** and **2**) and 80–98% in the outdoor ambient air, respectively. The unaccounted mass can be explained by the presence of oxide species, heteroatoms from the carbonaceous compounds and some water molecules (moisture, formation and crystallisation water) that have not been determined. The relative chemical composition of $PM_{2.5}$ was markedly different between subway platform and outdoor ambient air due to distinct emission source contributions, whereas the distributions of the chemical components were similar in the three subway systems studied.

Iron was the most abundant element in $PM_{2.5}$ found in the subway stations, with relative contribution to the bulk $PM_{2.5}$ ranging from 19 to 46% (**Figures 1** and **2**). The considerable amount of Fe in the subway stations is mainly attributed to mechanical friction and wear processes at rail-wheel-brake interfaces [9, 18, 30]. However, wear and friction processes initially produce Fe metal particles, and the surface of the primary particles must be reactive enough with oxygen in the air to easily react on the metallic surface, resulting in the formation of iron oxides [5]. Previous studies have reported Fe as the most abundant species in other subway systems [11–13, 16, 17, 31, 32]. Furthermore, the relative abundance of Fe in $PM_{2.5}$ on the platforms in the Barcelona subway system during the warmer period (19–33%) was lower than that measured on the platforms in Nomismatokopio (36–46%) and Bolhão (27–45%) stations [8]. Given that all three of these subway systems use trains with metallic wheels, this lower relative abundance of Fe in $PM_{2.5}$ on the platforms of Barcelona could be attributable to the presence of strong forced ventilation in the subway system in the warmer period, since in the colder period the Fe abundance was similar (27–46%) to that in Oporto and Athens subway systems. By comparison, aerosol samples collected outdoors contained less than 1% of Fe mass concentration.

The second largest chemical component of the subway $PM_{2.5}$ samples was that of total carbon, with mean relative contributions of this important subgroup ranging from 9 to 26% [8]. By comparison, in the outdoor urban air, TC concentrations were generally lower, although their relative contribution was higher (accounting for 17–39% of $PM_{2.5}$) due to the lower bulk $PM_{2.5}$ concentrations (**Figures 1** and **2**). It is important to note that in the three subway systems all trains are powered by electricity; thus, there are no combustion sources of TC. However, in Barcelona and Athens the TC concentrations on the platforms were around three times higher than those in the associated outdoor ambient air (**Figure 1b**). Possible sources of this TC are diesel-powered trains used for maintenance activities running at night, and especially the abrasion of C-bearing brakes and catenary power supply materials [8, 9, 33]. By contrast, in Oporto the TC concentrations were very similar between the platform and the outdoor air, indicating the clear influence of outdoor air in the Bolhão station which is followed in the line by an above ground station. Hence, the carbonaceous particles on the platform can arise from the outdoor environment in addition to those generated inside.

Elements of crustal origin were found in higher concentrations in subway $PM_{2.5}$ samples in comparison to outdoor ambient air, with relative contributions of crustal matter in the range of 5–12%, representing the third most abundant chemical component on the subway platforms (**Figures 1** and **2**). CM is present in outdoor $PM_{2.5}$ samples, as it derives from soil and urban mineral dust, although in $PM_{2.5}$ a low contribution is expected given the dominantly coarse mode of mineral matter. It is probable that some of these inhalable mineral particles measured on the subway platforms were brought in from the outdoor environment by the commuting passengers and by air exchange between the indoor and outdoor environments.

In addition, however, an important local source of crustal particles present on subway platforms will result from the resuspension of particles generated by weathering and erosion of ballast and construction materials, sometimes enhanced by occasional construction works in the subway systems.

Secondary inorganic compounds accounted for 2–10% of the total $PM_{2.5}$ subway concentrations. In general, SIC are one of the most abundant components in the outdoor atmosphere, accounting for 19–39% of the total $PM_{2.5}$ (**Figures 1** and **2**), indicating that these particles in the subway environment might arise from the outdoor environment. Moreover, the highest ws-SO_4^{2-} concentrations were recorded in the warmer period and the highest ws-NO_3^- were recorded in the colder period in Barcelona, according to the outdoor concentrations, which have a similar seasonal variation [34]. The relative amount of SIC in the total $PM_{2.5}$ was higher (10%) in the new station, given that the indoor sources for this station were lower.

The halite present in the subway environment is expected to come from outdoors by both air and water infiltration, the latter related to the evaporation of water and subsequent resuspension of halite minerals. The concentrations of NaCl were broadly similar at both Barcelona and Athens stations, and comparable to concentrations outdoors. By contrast, Oporto halite concentrations were notably higher in both the subway environment and outdoors, reflecting the Atlantic location of the city (**Figure 1b**).

Mean concentrations of insoluble sulphate ranged between 0.1 and 1.0 μg m^{-3} associated with the use of barite (highly water insoluble, $BaSO_4$) as a bulk material (as mineral filler) in the fabrication of brakes in trains [11].

In addition to the characteristically high Fe loading of subway particles, several other metals showed enhanced levels as compared to normal concentrations outdoors. These subway-enhanced trace metals included Ba, Cu, Mn, Zn, Cr, Sb, Sr, Zr, Ni, Sn, As and Co, indicating the presence of metal particle sources specific to the underground system (**Figure 5**). As expected, in the Barcelona study the trace elements concentrations in the colder period were higher than those in the warmer period due to the different ventilation programmes, as previously stated. Among all the studied stations of the three subway systems, the lowest concentration of trace elements was observed in the new Llefià station. Although these trace elements comprise less than 2% of the total $PM_{2.5}$, they are potentially useful for source identification, given the fact that they likely relate to chemical differences between rails and wheels (Mn, Cr), brakes (Ba, Sb, Cu, Zn, Pb, Ni, Sr) and power supply materials (e.g. Cu-rich catenaries and Cu vs. C pantographs) [9, 33]. Metalliferous particles released from these materials originate from mechanical wear and friction processes, as reported by other studies in subway systems [15, 31, 35, 36]. Improved ventilation, perhaps combined with changes in the mix of trace metals used in these subway materials, would potentially reduce commuter exposure to the trace metals.

Other trace elements identified in the chemical analyses (Pb, V, Li, Zr, Se, Rb, Y, Cd, La, Ce, Pr, Nd, Hf, Bi and U) together comprise a negligible amount (<0.1%) of the total $PM_{2.5}$. Such elements are not especially characteristic of the subway environment, being present in similar concentrations both below and above ground, and are probably associated with the infiltration of ambient city air in the subway systems.

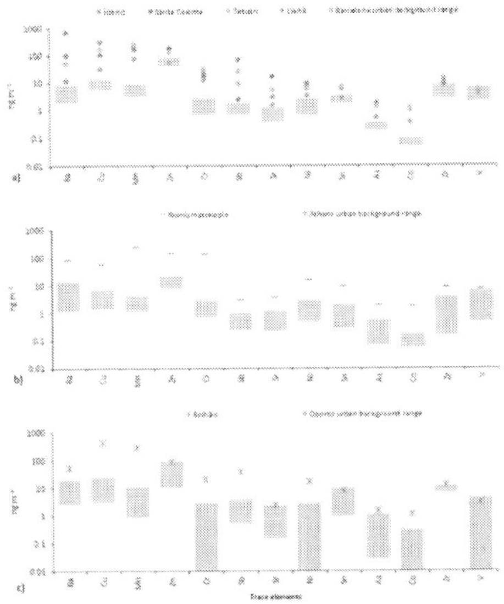

Figure 5. The mean concentration of trace elements (ng m^{-3}) on the platforms and the simultaneous urban background range (ambient site) in Barcelona **(a)**, Athens **(b)** and Oporto **(c)**.

7. Source contributions

For the Barcelona subway study, the number of $PM_{2.5}$ sources identified by PMF analysis varied from one station to another, but they can be grouped into outdoor and subway sources, the latter including all emissions generated by the circulation of trains (rails, wheels, brakes, catenaries and pantographs). The main differences among stations are attributed to (i) the different characteristics for each station, leading to different influences of the subway emissions on the platform environment, (ii) the different chemical composition of rails, wheels, brakes and power supply materials and (iii) the different influence of outdoor air, which is affected by the time of the year, among other factors.

The outdoor $PM_{2.5}$ sources found in the subway environment included secondary aerosol, sea salt and fuel-oil combustion. In the warmer period, the secondary source was characterised by a high contribution of sulphate, whereas in the colder it was dominated by nitrate. The sea salt source (mainly characterised by the presence of Na and Cl) had similar contributions in the warmer and colder periods. Moreover, a source characterised by V was identified, representing fuel-oil combustion [37].

The identified subway source had a different chemical profile for each of the stations. Although it was dominated by Fe at all stations, its content varies among stations (53–68%). Moreover, the relative abundance of other specific elements (e.g. Cu, Ba and Sr) of the subway $PM_{2.5}$ varied from station to station [9]. This source is identified as including over half of the Al_2O_3, Ca, Fe, Cr, Mn, Cu, Sr, Ba, Pr and Nd concentrations measured in all stations sampled (as well as over half of the Mg, Li, Ti, Co, Zn and Ce measured in the older stations of Joanic, Tetuan and Santa Coloma) [9].

The subway contribution was much lower during the warmer period (9–29%) than during the colder period (32–58%), this being attributed to the different ventilation, which allows for a better dispersion of the subway emissions in the warmer period.

8. Conclusions

This work is based on a large dataset from intensive and extensive measurement campaigns, aiming to characterise the air quality in terms of PM in three European subway systems (Barcelona, Athens and Oporto), both on platforms and inside trains.

There are important factors influencing PM concentrations in the subway systems, such as differences in the design of the stations and tunnels, rails geometry (curved vs. straight and sloped vs. levelled), system age, train frequency, ventilation and air-conditioning systems, passenger densities and outdoor air quality.

PM concentrations in subway platforms display clear diurnal patterns driven by the train frequency and the ventilation settings, with higher concentrations during subway operating hours. Moreover, in some cases the PM_X concentrations show temporal and spatial variations along the platforms, influenced in addition to the ventilation settings, by the design of the stations and tunnels, location of passengers' access to the platforms, commuter densities, and the effect of the passage and frequency of the trains.

PM_X concentrations inside the trains are very dependent on air-conditioning system, windows open/close, travelling above/underground, and PM_X concentrations on platforms and tunnels, showing short-time variations when doors open.

Subway aerosol is a complex mixture of components including iron, total carbon, crustal matter, secondary inorganic compounds, insoluble sulphate, halite and trace elements. Subway $PM_{2.5}$ is characterised by high concentrations of Fe (relative contribution to the bulk $PM_{2.5}$ ranging from 19 to 46%) in the three subway systems studied, generated mainly from mechanical wear and friction processes at rail-wheel-brake interfaces. Other trace elements with high enrichment in the subway $PM_{2.5}$ are Ba, Cu, Mn, Zn, Cr, Sb, Sr, Ni, Sn, As, Co and Zr. All metals present in the alloys used in the production of rails, wheels, brakes and power supply materials clearly suggest the wear of metal parts as the most important $PM_{2.5}$ subway source. In addition to the subway source, the contributions of secondary aerosol, sea salt and fuel-oil combustion sources can also be quantified.

http://dx.doi.org/10.5772/65364

This work expects to serve as a tool to establish the actions towards an effective control and to improve the air quality in subway systems, by identifying and encouraging the application of practical and focused air pollution mitigation strategies, appropriate for subway systems.

Acknowledgements

The present study was supported by the European Union Seventh Framework Programme (FP7/2007-2013) under Grant agreement no. 315760 HEXACOMM, the Spanish Ministry of Economy and Competitiveness and FEDER funds (METRO CGL2012-33066), the IMPROVE LIFE (LIFE13 ENV/ES/000263) and the AGAUR 2014 SGR33 grant from the Generalitat de Catalunya. Special thanks are also mentioned to Transports Metropolitans de Barcelona, the Urban Rail Transport S.A. (Athens Metro) and the Metro do Porto S.A.

Author details

Vânia Martins[1], María Cruz Minguillón[1], Teresa Moreno[1*], Luís Mendes[2], Konstantinos Eleftheriadis[2], Célia A. Alves[3], Eladio de Miguel[4] and Xavier Querol[1]

*Address all correspondence to: teresa.moreno@idaea.csic.es

1 Institute of Environmental Assessment and Water Research (IDAEA), CSIC, Barcelona, Spain

2 Environmental Radioactivity Laboratory, Institute of Nuclear & Radiological Sciences & Technology, Energy & Safety, N.C.S.R. 'Demokritos', Athens, Greece

3 Department of Environment, Centre for Environmental and Marine Studies (CESAM), University of Aveiro, Portugal

4 Transports Metropolitans de Barcelona (TMB), L'Hospitalet de Llobregat, Spain

References

[1] Bachoual R, Boczkowski J, Goven D, et al. Biological effects of particles from the Paris subway system. *Chem Res Toxicol* 2007; 20: 1426–1433.

[2] Bigert C, Alderling M, Svartengren M, et al. Blood markers of inflammation and coagulation and exposure to airborne particles in employees in the Stockholm underground. *Occup Environ Med* 2008; 65: 655–658.

[3] Salma I, Pósfai M, Kovács K, et al. Properties and sources of individual particles and some chemical species in the aerosol of a metropolitan underground railway station. *Atmos Environ* 2009; 43: 3460–3466.

[4] Nieuwenhuijsen MJ, Gómez-Perales JE, Colvile RN. Levels of particulate air pollution, its elemental composition, determinants and health effects in metro systems. *Atmos Environ* 2007; 41: 7995–8006.

[5] Jung H-J, Kim B, Ryu J, et al. Source identification of particulate matter collected at underground subway stations in Seoul, Korea using quantitative single-particle analysis. *Atmos Environ* 2010; 44: 2287–2293.

[6] Loxham M, Cooper MJ, Gerlofs-Nijland ME, et al. Physicochemical characterization of airborne particulate matter at a mainline underground railway station. *Environ Sci Technol* 2013; 47: 3614–3622.

[7] Sundh J, Olofsson U, Olander L, et al. Wear rate testing in relation to airborne particles generated in a wheel – rail contact. *Lubr Sci* 2009; 21: 135–150.

[8] Martins V, Moreno T, Mendes L, et al. Factors controlling air quality in different European subway systems. *Environ Res* 2016; 146: 35–46.

[9] Martins V, Moreno T, Minguillón MC, et al. Origin of inorganic and organic components of PM2.5 in subway stations of Barcelona, Spain. *Environ Pollut* 2016; 208: 125–136.

[10] Zimmer AT, Maynard AD. Investigation of the aerosols produced by a high-speed, hand-held grinder using various substrates. *Ann Occup Hyg* 2002; 46: 663–672.

[11] Aarnio P, Yli-Tuomi T, Kousa A, et al. The concentrations and composition of and exposure to fine particles (PM2.5) in the Helsinki subway system. *Atmos Environ* 2005; 39: 5059–5066.

[12] Chillrud SN, Epstein D, Ross JM, et al. Elevated airborne exposures to manganese, chromium and iron of teenagers from steel dust and New York City's subway system. *Environ Sci Technol* 2004; 38: 732–737.

[13] Mugica-Álvarez V, Figueroa-Lara J, Romero-Romo M, et al. Concentrations and properties of airborne particles in the Mexico City subway system. *Atmos Environ* 2012; 49: 284–293.

[14] Murruni LG, Solanes V, Debray M, et al. Concentrations and elemental composition of particulate matter in the Buenos Aires underground system. *Atmos Environ* 2009; 43: 4577–4583.

[15] Querol X, Moreno T, Karanasiou A, et al. Variability of levels and composition of PM10 and PM2.5 in the Barcelona metro system. *Atmos Chem Phys* 2012; 12: 5055–5076.

[16] Salma I, Weidinger T, Maenhaut W. Time-resolved mass concentration, composition and sources of aerosol particles in a metropolitan underground railway station. *Atmos Environ* 2007; 41: 8391–8405.

[17] Johansson C, Johansson P-A. Particulate matter in the underground of Stockholm. *Atmos Environ* 2003; 37: 3–9.

[18] Kam W, Delfino RJ, Schauer JJ, et al. A comparative assessment of PM2.5 exposures in light-rail, subway, freeway, and surface street environments in Los Angeles and estimated lung cancer risk. *Environ Sci Process Impacts* 2013; 15: 234–243.

[19] Guo L, Hu Y, Hu Q, et al. Characteristics and chemical compositions of particulate matter collected at the selected metro stations of Shanghai, China. *Sci Total Environ* 2014; 496: 443–452.

[20] Kwon S-B, Jeong W, Park D, et al. A multivariate study for characterizing particulate matter (PM10, PM2.5, and PM1) in Seoul metropolitan subway stations, Korea. *J Hazard Mater* 2015; 297: 295–303.

[21] Moreno T, Pérez N, Reche C, et al. Subway platform air quality: Assessing the influences of tunnel ventilation, train piston effect and station design. *Atmos Environ* 2014; 92: 461–468.

[22] Park D-U, Ha K-C. Characteristics of PM10, PM2.5, CO2 and CO monitored in interiors and platforms of subway train in Seoul, Korea. *Environ Int* 2008; 34: 629–34.

[23] Ripanucci G, Grana M, Vicentini L, et al. Dust in the underground railway tunnels of an Italian town. *J Occup Environ Hyg* 2006; 3: 16–25.

[24] Lim J-M, Lee J-H, Moon J-H, et al. Source apportionment of PM10 at a small industrial area using positive matrix factorization. *Atmos Res* 2010; 95: 88–100.

[25] Park D, Lee T, Hwang D, et al. Identification of the sources of PM10 in a subway tunnel using positive matrix factorization. *J Air Waste Manage Assoc* 2014; 64: 1361–1368.

[26] Escrig A, Monfort E, Celades I, et al. Application of optimally scaled target factor analysis for assessing source contribution of ambient PM10. *J Air Waste Manage Assoc* 2009; 59: 1296–1307.

[27] Paatero P, Tapper U. Positive matrix factorization: A non-negative factor model with optimal utilization of error estimates of data values. *Environmetrics* 1994; 5: 111–126.

[28] Martins V, Moreno T, Minguillón MC, et al. Exposure to airborne particulate matter in the subway system. *Sci Total Environ* 2015; 511: 711–722.

[29] Raut J-C, Chazette P, Fortain A. Link between aerosol optical, microphysical and chemical measurements in an underground railway station in Paris. *Atmos Environ* 2009; 43: 860–868.

[30] Park D, Oh M, Yoon Y, et al. Source identification of PM10 pollution in subway passenger cabins using positive matrix factorization. *Atmos Environ* 2012; 49: 180–185.

[31] Furuya K, Kudo Y, Okinaga K, et al. Seasonal variation and their characterization of suspended particulate matter in the air of subway stations. *J Trace Microprobe Tech* 2001; 19: 469–485.

[32] Seaton A, Cherrie J, Dennekamp M, et al. The London Underground: dust and hazards to health. *Occup Environ Med* 2005; 62: 355–62.

[33] Moreno T, Martins V, Querol X, et al. A new look at inhalable metalliferous airborne particles on rail subway platforms. *Sci Total Environ* 2015; 505: 367–375.

[34] Querol X, Alastuey A, Moreno T, et al. Spatial and temporal variations in airborne particulate matter (PM10 and PM2.5) across Spain 1999–2005. *Atmos Environ* 2008; 42: 3964–3979.

[35] Cheng Y-H, Lin Y-L, Liu C-C. Levels of PM10 and PM2.5 in Taipei rapid transit system. *Atmos Environ* 2008; 42: 7242–7249.

[36] Gustafsson M, Blomqvist G, Swietlicki E, et al. Inhalable railroad particles at ground level and subterranean stations—Physical and chemical properties and relation to train traffic. *Transp Res Part D Transp Environ* 2012; 17: 277–285.

[37] Agrawal H, Malloy QGJ, Welch WA, et al. In-use gaseous and particulate matter emissions from a modern ocean going container vessel. *Atmos Environ* 2008; 42: 5504–5510.

Chapter 6

The Effects of Urban Public Transit Investment on Traffic Congestion and Air Quality

Justin Beaudoin and C.-Y. Cynthia Lin Lawell

Additional information is available at the end of the chapter

http://dx.doi.org/10.5772/66834

Abstract

Traffic congestion is ubiquitous across urban roadways, and the adverse health effects accompanying deteriorating air quality are an ongoing concern. Beyond these local effects, transportation is also a major contributor of greenhouse gas emissions and is thus a significant element of the climate change debate. A contentious issue currently confronting transportation analysts and policy-makers is what the effects of public transit investment on traffic congestion and on air quality are and therefore what the appropriate level of public transit investment should be. While public transit receives plenty of political support for its "green" reputation and its contribution to sustainability, there have been relatively few studies examining the ex post–effects of public transit investment on traffic congestion or air quality. In this chapter, we review our theoretical and empirical research on the effects of public transit investment on congestion, the demand for automobile travel, and air quality.

Keywords: public transit, congestion, air quality

1. Introduction

Traffic congestion is ubiquitous across urban roadways and the adverse health effects accompanying deteriorating air quality are an ongoing concern. Beyond these local effects, transportation is also a major contributor of greenhouse gas emissions and is thus a significant element of the climate change debate.

The Texas Transportation Institute made headlines with its estimate of the annual costs of traffic congestion in the USA exceeding $120 billion in 2011 [1], owing primarily to the costs imposed by excessive traffic levels on travel times for freight and personal travel. Studies have shown that traffic congestion is the number one concern of individuals in rapidly growing areas in the USA, often ranked higher than crime, school overcrowding, and housing shortages [2].

Urban transportation not only leads to traffic congestion but also to air pollution. In 2010, vehicle emissions in the USA contributed to an estimated 2200 premature deaths and more than $18 billion in public health expenditures [3].

It is clear that the market failures endemic to the urban transportation sector are not being adequately addressed by existing regulatory policies [4]. The market failures in the auto and transit sectors have long been of interest [4].

The government has two potential roles in the surface transportation sector. The first potential role for government is to provide transportation infrastructure in the form of roads and public transit systems and also to operate public transit services. Once the infrastructure is in place, the second potential role for government is to employ policy instruments (such as taxes and other forms of regulation relating to safety, environmental standards, travel demand management policies, and so forth) in order to address the market failures that are inherent to unregulated transportation activity and also to determine the operational aspects of public transit service [4, 5].

A contentious issue currently confronting transportation analysts and policy-makers is what the effects of public transit investment on congestion and on air quality are and therefore what the appropriate level of public transit investment should be [4]. While public transit receives plenty of political support for its "green" reputation and its contribution to sustainability, there have been relatively few studies examining the ex post effects of public transit investment on traffic congestion or air quality [4].

For example, previous empirical studies examining the relationship between transit supply and traffic congestion are limited, and the findings of these studies vary [4]. There is also an ongoing debate in policy circles regarding the efficacy of public transit investment as a means of addressing traffic congestion, for example [6–8], all display skepticism regarding the congestion-reduction possibilities of public transit, while [9] advocates for transit investment [4]. Although investment in public transit may lead to short-term reductions in congestion due to a "substitution effect," in the long run, it may be less effective due to the "induced demand effect" [4, 10, 11].

Similarly, while several studies have considered the relationship between automobile travel and air quality, there have been relatively few empirical studies looking at the effect of public transit on air quality [4]. Although there is generally a consensus that auto travel leads to adverse health outcomes, there is very little empirical evidence on the incremental effect that public transit supply may or may not have on air quality [4].

With $18 billion spent on public transit capital in the USA each year [12], it is imperative to assess the effects of these expenditures on transportation activity and the environment and what path future investment should take [4].

In this chapter, we review our theoretical and empirical research on the effects of public transit investment on congestion, the demand for automobile travel, and air quality. In Ref. [5], we develop a theory model to evaluate whether public transit investment has a role in reducing congestion a second-best setting. In Refs. [11, 13], we empirically analyze the effects of public transit investment on the demand for automobile travel and on air pollution, respectively, by applying an instrumental variable approach that accounts for the potential endogeneity of public transit investment to a uniquely created panel dataset of 96 urban areas across the USA over the years 1991–2011.

Our results in Ref. [5] suggest that investments in public transit may have a co-benefit of congestion reduction. Thus, when analyzing potential public transit projects using a cost-benefit analysis framework, interactions between auto and transit users should be taken into account. However, while public transit investment may be able to play a complementary role, efficient pricing of auto travel remains necessary to address traffic congestion in the USA [5].

Our results in Ref. [11] show that, owing to the substitution effect, increases in public transit supply lead to a reduction in the demand for automobile travel, but that this reduction can be offset at least in part by induced demand. Moreover, the magnitude of the effect of public transit on the demand for automobile travel is subject to heterogeneity across urban areas. We also find in Ref. [11] that, for both the substitution effect and the equilibrium effect (which incorporates both the substitution effect and induced demand), public transit supply does not reduce the demand for automobile travel until the demand for automobile travel exceeds a minimum threshold and that beyond this threshold the magnitude of the negative elasticity of the demand for automobile travel with respect to transit capacity increases with the demand for automobile travel [11].

In Ref. [13], we analyze the effects of the level of transit supply on ambient concentrations of carbon monoxide, lead, nitrogen dioxide, ozone, particulate matter, and sulfur dioxide. We find that—at the margin and given existing urban travel regulations in place—there is no evidence that an increase in transit supply improves air quality [13].

Our research in Refs. [11, 13] improves upon previous empirical studies by using a broader set of urban areas over a longer time period than previous studies and by allowing for heterogeneity in the effects.

Our results have important implications for the design of sustainable transportation policy and for urban transport systems and are of interest to academics and policy-makers alike.

2. Literature review

In this section we review the related literature. We provide a thorough and detailed literature review in Ref. [4].

2.1. Congestion

When externalities arising from auto travel are not internalized, these auto market distortions may have implications for the optimal level of public transit investment. If the speed of auto travel is independent of the volume of transit service, then the second-best transit capacity that accounts for the distortion in the auto market is higher than the first-best transit capacity that would be provided if there were no auto market distortions that needed to be accounted for (e.g., if the auto market distortions were already separately addressed via a Pigouvian tax) [4, 5, 14]. However, if auto and transit modes are interdependent, then the optimal level of transit capacity to provide depends on the extent to which public transit affects the deadweight loss associated with auto travel [4, 5, 15–17].

According to the "fundamental law of traffic congestion," while investment in infrastructure may lead to short-term reductions in congestion, in the long run, it will be ineffective in the absence of efficient pricing due to "induced demand" (often referred to as "latent demand") [4, 5, 11, 18].

The existing empirical evidence on the effect of public transit on traffic congestion is unclear [4]. In their analysis of the effects of rail transit investment on the share of public transit ridership [19], find that, consistent with the "fundamental law of traffic congestion" and the presence of induced demand, rail transit investment does not reduce congestion levels, though it does lead to reduced commuting times for the subset of commuters that switch from bus to rail [4, 19].

In their study of the effects of roadway expenditures on the cost of congestion [10], find that increases in the mileage of rail transit lead to a reduction in congestion costs, but that increases in bus service actually increase congestion costs. Their results are consistent with a congestion externality interdependency between auto and transit travel [4, 10].

In their test of the "fundamental law of traffic congestion" [20], find compelling evidence of induced auto demand, with increases in road capacity being met with commensurate increases in auto travel [4, 20].

In his analysis of travel speeds before and during a transit labor dispute within the Los Angeles transit system in 2003 [21], finds that the average highway delay increased by 47% when transit service ceased operation. The results of Refs. [21–23] provide convincing evidence of the effects of transit at the extensive margin when compared with the counterfactual scenario of no transit service [4].

In their study of the impact of bicycle-sharing infrastructure on urban transportation [24], find that the availability of a bikeshare reduces traffic congestion over 2–3% within a neighborhood. They also find that the congestion-reducing impact of bikeshares is concentrated in highly congested areas [24].

The effects of public transportation and the built environment on the number of civilian vehicles in China are analyzed by Liu and Lin Lawell [25], who apply a two-step Generalized Method of Moments (GMM) instrumental variable model to city-level panel data over the period 2001–2011. The results show that increasing the road area increases the number of civilian vehicles, which provides empirical support for the "fundamental law of traffic congestion" in China. In contrast, increasing the public transit passenger load decreases the number of civilian vehicles, suggesting that public transportation and civilian cars are substitutes. The effects vary by city population, however. For larger cities, increases in the number of public buses increase the number of civilian vehicles, but increases in the number of taxis and in road area decrease the number of civilian vehicles. They also find that land use diversity increases the number of civilian vehicles, especially in the higher-income cities and in the extremely big cities. There is no significant relationship between civilian vehicles and per capita disposable income except in mega cities [25].

Overall, the existing empirical evidence of the effect of transit investment on traffic congestion is mixed [4, 11]. The conflicting conclusions of previous studies may also be due to differences in empirical methodologies employed and the characteristics of the dataset used. Our work in Ref. [11] uses a broader set of urban areas over a longer time period than previous studies, and the regional heterogeneity that our results in Ref. [11] indicate helps to reconcile the seemingly conflicting evidence from the previous literature.

2.2. Air quality

While several studies have considered the relationship between auto travel and air quality [3, 26–28], and the effects of transportation policies such as driving restrictions on air quality [29, 30], there have been relatively few empirical studies looking at the effect of public transit on air quality [4].

In particular, while there is generally a consensus that auto travel leads to adverse health outcomes, there is very little empirical evidence of the incremental effect that transit supply may or may not have on air quality [4]. Two recent studies have provided an initial look at the relationship between transit supply and air quality. Using hourly air quality data from Taipei [31], find that the new rail system's opening reduced carbon monoxide by 5–15% but had little effect on ground level ozone pollution [4, 31]. In their analysis of the environmental effect of expanded rail service in Germany over the period 1994–2004 [32], find that increases in rail service frequency lead to a reduction in some pollutants (NO, NO_2, and CO), though not others (SO_2 and O_3) [4, 32].

The effects of public transit on air quality depend on the relative demand substitutability between auto and transit, and on the extent to which the emission rates vary between auto and transit travel, and are therefore an empirical issue [4]. As the relationship between transit and observed pollution levels is theoretically ambiguous, it is difficult to impute the effect of transit on air quality based on previous studies that focus on the effects of auto travel on air quality [4, 33].

3. Results

In this section we review our theoretical and empirical research on the effects of public transit investment on congestion, the demand for automobile travel, and air quality.

3.1. Theory of public transit investment and traffic congestion policy

In Ref. [5], we develop a theory model to evaluate whether public transit investment has a role in reducing congestion a second-best setting. The model enables us to evaluate the extent to which traffic congestion should be accounted for when evaluating investment in public transit infrastructure when a Pigouvian congestion tax cannot be levied on auto travel. We contribute to the literature by allowing for both demand and cost interdependencies across the auto and transit modes. We find that the level of transit investment should be higher relative to that chosen when the congestion-reduction effects of transit are not accounted for. The importance of accounting for the congestion-reduction effects of transit depends upon the demand and cost interdependencies across the auto and transit modes, which may vary across regions [5].

Our results in Ref. [5] suggest that investments in public transit may have a co-benefit of congestion reduction. Thus, when analyzing potential public transit projects using a cost-benefit analysis framework, interactions between auto and transit users should be taken into account. However, while public transit investment may be able to play a complementary role, efficient pricing of auto travel remains necessary to address traffic congestion in the USA [5].

3.2. The effects of public transit supply on the demand for automobile travel

On average, the total hours of delay attributable to congestion in urban areas in the USA have more than tripled over the past three decades, during which there has been an 83% increase in auto travel, a 16% increase in transit travel, and a 16% increase in travel times [11].

Over the last two decades, the volume of public transit travel in the USA has increased by 43% [11]. During this period, the overall transit network coverage (directional route miles) has increased by approximately 35%, while the capacity provided over the network (vehicle miles per directional route mile) has increased by approximately 11%, yielding an overall increase in total vehicle miles supplied by public transit of 50% [11].

Although investment in public transit may lead to short-term reductions in congestion due to a "substitution effect," in the long run, it may be less effective due to the "induced demand effect" [10, 11]. In Ref. [11] we empirically analyze the effects of public transit investment on the demand for automobile travel by applying an instrumental variable approach that accounts for the potential endogeneity of public transit investment to a uniquely created panel dataset of 96 urban areas across the USA over the years 1991–2011. We estimate both the short-run substitution effect and the longer-run equilibrium effect that account for both the substitution effect and the induced demand effect [11].

To estimate the substitution effect, we run the following regression [11]:

$$autotravel_{rt} = \beta_1 transit_{rt} + x_{rt}'\beta_2 + \alpha_r + \varepsilon_{rt}, \tag{1}$$

where $autotravel_{rt}$ is the demand for auto travel in region r in year t, as measured by the number of vehicle miles traveled per freeway lane mile; $transit_{rt}$ is the public transit supply in region r in year t, as measured by vehicle revenue miles; x_{rt} is a vector of control variables in region r in year t, including freeway capacity, arterial road capacity, fuel cost, transit fare, employment, income, population, year, and year squared; and α_{rt} is a region fixed effect. We use instruments to address the endogeneity of public transit supply $transit_{rt}$ [11].

To estimate the equilibrium effect accounting for both the substitution effect and the induced demand effect, we remove the factors associated with the induced demand effect (employment, income, and population) and instead control for their initial levels in the base year of 1991. In particular, we run the following regression [11]:

$$autotravel_{rt} = \beta_1 transit_{rt} + x_{rt}' \beta_2 + \varepsilon_{rt}, \tag{2}$$

where $autotravel_{rt}$ is the demand for auto travel in region r in year t, as measured by the number of vehicle miles traveled per freeway lane mile; $transit_{rt}$ is the public transit supply in region r in year t, as measured by vehicle revenue miles; and x_{rt} is a vector of control variables for in region r in year t, including freeway capacity, arterial road capacity, fuel cost, transit fare, employment in the base year 1991, income in the base year 1991, population in the base year 1991, year, and year squared. We use instruments to address the endogeneity of public transit supply $transit_{rt}$ [11].

To address the potential endogeneity of public transit investment, we use two sources of instrumental variables for public transit investment in our analyses in Ref. [11]. The first instrument we use is lagged political voting records. In particular, we use as instruments the Democratic voting share within the urban area averaged over lagged Presidential, Gubernatorial, or Senate elections [11, 13]. Democratic voters are much more likely than Republican voters to support referenda in relation to public transit investment [34]. Democratic voting shares are expected to be related to public transit investment through two channels: (1) through the effect on the total public funds budget and (2) through relatively stronger preferences for public transit and thus the allocation of total public funds directed to public transit [11].

Conditional on time-invariant region-specific factors that are absorbed by the regional fixed effects, changes in lagged voting records are not related to congestion except through their effect on public transit investment. Similarly, after controlling for employment rate, income, and population, factors causing changes in the lagged Democratic voting share within the urban area in Presidential, Gubernatorial, or Senate elections are unlikely to be related to factors that are causing changes in local congestion, as congestion is not an issue that influences elections above the local level. After conditioning on these variables, voting records can be interpreted as a proxy for underlying transit preferences in the region that is orthogonal to congestion [11].

The second instrument we use for public transit investment in our empirical analyses in Ref. [11] is the lagged level of Federal funds provided for transit in the region. While Local and State funds may be correlated with unobserved factors affecting regional congestion, conditional on time-invariant region-specific unobservables that are absorbed by the regional fixed effects, changes in lagged Federal funds are orthogonal to such potential factors [11].

The data we use in Ref. [11] covers 96 urban areas within 351 counties and 44 states across the USA and spans 21 years from 1991 to 2011. As defined by the Census Bureau, an "urban area" (UZA) refers to a region that is centered around a core metropolitan statistical area (MSA). The data we use in Ref. [11] relating to the auto travel components of each UZA's transportation network are primarily from the Texas Transportation Institute's Urban Mobility Report [1], which are the "best available means of comparing congestion levels in different regions and tracking changes in regional congestion levels over time" ([35], p. 17). The Urban Mobility Report measures traffic delay using data from the US Department of Transportation on traffic volumes and the characteristics of the city (see Ref. [10], p. 467 for discussion). While we measure congestion as the daily vehicle miles traveled per freeway lane mile, our empirical results in Ref. [11] are robust to the particular measure of congestion used.

Our results in Ref. [11] show that, owing to the substitution effect, increases in public transit supply lead to a reduction in the demand for automobile travel, but this reduction can be offset at least in part by induced demand. Moreover, the magnitude of the effect of public transit on the demand for automobile travel is subject to heterogeneity across urban areas. We also find in Ref. [11] that, for both the substitution effect and the equilibrium effect (which incorporates both the substitution effect and induced demand), public transit supply does not reduce the demand for automobile travel until the demand for automobile travel exceeds a minimum threshold and that beyond this threshold the magnitude of the negative elasticity of the demand for automobile travel with respect to transit capacity increases with the demand for automobile travel [11].

Our research in Ref. [11] improves upon previous empirical studies by using a broader set of urban areas over a longer time period than previous studies and by allowing for heterogeneity in the effects.

3.3. Evaluating the effects of transit supply on air quality

In Ref. [13], we analyze the effects of the level of transit supply on ambient concentrations of carbon monoxide, lead, nitrogen dioxide, ozone, particulate matter, and sulfur dioxide.

In particular, we empirically analyze the effects of public transit investment on air pollution by applying an instrumental variable approach that accounts for the potential endogeneity of public transit investment to a uniquely created panel dataset of 96 urban areas across the USA over the years 1991–2011 [13].

We run the following regression for each air pollutant (CO, NO_2, O_3, PM_{10}, $PM_{2.5}$, and SO_2) [13]:

$$pollution_{rt} = \beta_1 \, transit_{rt} + x_{rt}' \beta_2 + \alpha_r + \varepsilon_{rt}, \tag{3}$$

where $pollution_{rt}$ is the ambient level of that air pollutant in region r in year t; $transit_{rt}$ is the public transit supply in region r in year t, as measured by vehicle revenue miles; x_{rt} is a vector of control variables for in region r in year t, including freeway capacity, arterial road capacity, fuel cost, transit fare, trucking activity, employment, income, population, pollution point sources, weather controls, and dummies for National Ambient Air Quality Standards; and α_{rt} is a region fixed effect. We use instruments to address the endogeneity of public transit supply $transit_{rt}$ [13].

To address the potential endogeneity of public transit investment, we use two sources of instrumental variables for public transit investment in our analyses in Ref. [13]. The first instrument we use is lagged political voting records. In particular, we use as instruments the Democratic voting share within the urban area averaged over lagged Presidential, Gubernatorial, or Senate elections [11, 13]. Democratic voters are much more likely than Republican voters to support referenda in relation to public transit investment [34]. Democratic voting shares are expected to be related to public transit investment through two channels: (1) through the effect on the total public funds budget and (2) through relatively stronger preferences for public transit and thus the allocation of total public funds directed to public transit [11, 13].

Conditional on time-invariant region-specific factors that are absorbed by the regional fixed effects, changes in lagged voting records are not related to air quality except through their effect on public transit investment. Similarly, after controlling for employment rate, income, and population, factors causing changes in the lagged Democratic voting share within the urban area in Presidential, Gubernatorial, or Senate elections are unlikely to be related to factors that are causing changes in local air pollution, as air pollution is not an issue that influences elections above the local level. After conditioning on these variables, voting records can be interpreted as a proxy for underlying transit preferences in the region that is orthogonal to air quality [13].

The second instrument we use for public transit investment in our empirical analyses in Ref. [13] is the lagged level of Federal funds provided for transit in the region. While Local and State funds may be correlated with unobserved factors affecting regional air quality, conditional on time-invariant region-specific unobservables that are absorbed by the regional fixed effects, changes in lagged Federal funds are orthogonal to such potential factors [13].

The data we use in Ref. [13] covers 96 urban areas within 351 counties and 44 states across the USA and spans 21 years from 1991 to 2011. As defined by the Census Bureau, an "urban area" (UZA) refers to a region that is centered around a core metropolitan statistical area (MSA) [11, 13].

For the air quality data in Ref. [13], we use daily air quality data for each Core-Based Statistical Area (CBSA) recorded by the Environmental Protection Agency (EPA) at monitoring stations that measure the ambient level of CO, NO_2, O_3, PM_{10}, $PM_{2.5}$, and SO_2. Each CBSA is then mapped to the UZA of our dataset. On average, 98.6% of the UZA population is contained within the CBSA [13].

According to the results of our empirical analysis of the effects of the level of transit supply on ambient concentrations of carbon monoxide, lead, nitrogen dioxide, ozone, particulate matter, and sulfur dioxide in Ref. [13], we find that—at the margin and given existing urban travel regulations in place—there is no evidence that an increase in transit supply improves air quality [13].

Our research in Ref. [13] improves upon previous empirical studies by using a broader set of urban areas over a longer time period than previous studies.

4. Conclusion

It is clear that the market failures endemic to the urban transportation sector are not being adequately addressed by existing regulatory policies [4].

A contentious issue currently confronting transportation analysts and policy-makers is what the effects of public transit investment on traffic congestion and on air quality are and therefore what the appropriate level of public transit investment should be [4]. While public transit receives plenty of political support for its "green" reputation and its contribution to sustainability, there have been relatively few studies examining the ex post effects of public transit investment on traffic congestion or air quality [4].

With $18 billion spent on public transit capital in the USA each year [12], it is imperative to assess the effects of these expenditures on transportation activity and the environment and what path future investment should take [4].

In this chapter, we review our theoretical and empirical research on the effects of public transit investment on congestion, the demand for automobile travel, and air quality. In Ref. [5], we develop a theory model to evaluate whether public transit investment has a role in reducing congestion a second-best setting. In Refs. [11, 13], we empirically analyze the effects of public transit investment on the demand for automobile travel and on air pollution, respectively, by applying an instrumental variable approach that accounts for the potential endogeneity of public transit investment to a uniquely created panel dataset of 96 urban areas across the USA for the years 1991–2011.

Our results in Ref. [5] suggest that investments in public transit may have a co-benefit of congestion reduction. Thus, when analyzing potential public transit projects using a cost-benefit analysis framework, interactions between auto and transit users should be taken into account. However, while public transit investment may be able to play a complementary role, efficient pricing of auto travel remains necessary to address traffic congestion in the USA [5].

Our results in Ref. [11] show that, owing to the substitution effect, increases in public transit supply lead to a reduction in the demand for automobile travel, but this reduction can be offset at least in part by induced demand. Moreover, the magnitude of the effect of public transit on the demand for automobile travel is subject to heterogeneity across urban areas. We also find in Ref. [11] that, for both the substitution effect and the equilibrium effect (which incorporates both the substitution effect and induced demand), public transit supply does not reduce the demand for automobile travel until the demand for automobile travel exceeds a minimum threshold and that beyond this threshold the magnitude of the negative elasticity of the demand for automobile travel with respect to transit capacity increases with the demand for automobile travel [11].

In Ref. [13], we analyze the effects of the level of transit supply on ambient concentrations of carbon monoxide, lead, nitrogen dioxide, ozone, particulate matter, and sulfur dioxide. We find that—at the margin and given existing urban travel regulations in place—there is no evidence that an increase in transit supply improves air quality [13].

http://dx.doi.org/10.5772/66834

Our research in Refs. [11, 13] improves upon previous empirical studies by using a broader set of urban areas over a longer time period than previous studies and by allowing for heterogeneity in the effects.

Our results have important implications for the design of sustainable transportation policy and for urban transport systems and are of interest to academics and policy-makers alike.

Acknowledgements

We thank Brandon Schaufele, Y. Hossein Farzin, Erik Hurst, Doug Larson, Marcel Oesterich, Chad Sparber, and Matthew Turner; participants at the 2013 International Transportation Economics Association Conference, the 2013 Canadian Resource and Environmental Economists (CREE) Group Conference, and the 2016 Canadian Resource and Environmental Economists (CREE) Study Group Annual Conference; and participants in seminars at the University of California at Davis, California State University at Fullerton, Colgate University, the University of Washington at Tacoma, Furman University, and Western Washington University for helpful comments and discussions. Lin Lawell is a member of the Giannini Foundation of Agricultural Economics. All errors are our own.

Author details

Justin Beaudoin[1] and C.-Y. Cynthia Lin Lawell[2]*

*Address all correspondence to: cclin@primal.ucdavis.edu

1 University of Washington Tacoma, Tacoma, WA, USA

2 University of California at Davis, Davis, CA, USA

References

[1] Schrank D, Eisele B, Lomax T. Texas A&M Transportation Institute's 2012 Urban Mobility Report. College Station, TX: Texas Transportation Institute; 2012.

[2] Morrison GM, Lin Lawell C-YC. Does employment growth increase travel time to work?: An empirical analysis using military troop movements. Regional Science and Urban Economics. 2016;**60**:180–197.

[3] Levy J, Buonocore J, von Stackelberg K. Evaluation of the public health impacts of traffic congestion: A health risk assessment. Environmental Health. 2010;**9**:65–76.

[4] Beaudoin J, Farzin YH, Lin Lawell C-YC. Public transit investment and sustainable transportation: A review of studies of transit's impact on traffic congestion and air quality. Research in Transportation Economics. 2015;**52**:15–22.

[5] Beaudoin J, Farzin YH, Lin Lawell C-YC. Public transit investment and traffic congestion policy. Working paper, University of California at Davis in Davis, California; 2016.

[6] Rubin TA, Moore II JE, Lee S. Ten myths about US urban rail systems. Transport Policy. 1999;**6**:57–73.

[7] Stopher P. Reducing road congestion: A reality check. Transport Policy. 2004;**11**:117–131.

[8] Rubin TA, Mansour F. Transit Utilization and Traffic Congestion: Is There a Connection? Reason Foundation, Policy Study 427; 2013. Available from: http://reason.org/files/transit_utilization_traffic_congestion.pdf.

[9] Litman T. Critique of Transit Utilization and Traffic Congestion: Is There a Connection? Victoria Transport Policy Institute Report; 2014. Available from: www.vtpi.org/R&M_critique.pdf.

[10] Winston C, Langer A. The effect of government highway spending on road users' congestion costs. Journal of Urban Economics. 2006;**60**(3):463–483.

[11] Beaudoin J, Lin Lawell C-YC. The effects of public transit supply on the demand for automobile travel. Working paper, University of California at Davis in Davis, California; 2016.

[12] American Public Transportation Association. 2012 Public Transportation Fact Book. Washington, DC; 2012.

[13] Beaudoin J, Lin Lawell C-YC. Is public transit's "green" reputation deserved?: Evaluating the effects of transit supply on air quality. Working paper, University of California at Davis in Davis, California; 2016.

[14] Kraus M. Road pricing with optimal mass transit. Journal of Urban Economics. 2012;**72**:81–86.

[15] Ahn K. Road pricing and bus service policies. Journal of Transport Economics and Policy. 2009;**43**(1):25–53.

[16] Sherman R. Congestion interdependence and urban transit fares. Econometrica. 1971;**39**(3):565–576.

[17] Basso L, Silva H. Efficiency and substitutability of transit subsidies and other urban transport policies. American Economic Journal: Economic Policy. 2014;**6**(4):1–33.

[18] Hau T. Transport for urban development in Hong Kong. In: Transport and Communications for Urban Development: Report of the HABITAT II Global Workshop. Nairobi, Kenya: The United Nations Centre for Human Settlements (HABITAT); 1997. pp. 267–289.

[19] Baum-Snow N, Kahn ME. Effects of Urban Rail Transit Expansions: Evidence from Sixteen Cities. Brookings-Wharton Papers on Urban Affairs; 2005:147–206.

[20] Duranton G, Turner MA. The fundamental law of road congestion: Evidence from US cities. American Economic Review. 2011;**101**(6):2616–2652.

[21] Anderson ML. Subways, strikes, and slowdowns: The impacts of public transit on traffic congestion. American Economic Review. 2014;**104**(9):2763–2796.

[22] Winston C, Maheshri V. On the social desirability of urban rail transit systems. Journal of Urban Economics. 2007;**62**(2):362–382.

[23] Nelson P, Baglino A, Harrington W, Safirova E, Lipman A. Transit in Washington, DC: Current benefits and optimal level of provision. Journal of Urban Economics. 2007;**62**(2):231–251.

[24] Hamilton TL, Wichman CJ. Bicycle infrastructure and traffic congestion: Evidence from DC's Capital Bikeshare. Resources for the Future Discussion Paper, Washington, DC. 15–39; 2015.

[25] Liu Q, Lin Lawell C-YC. The effects of public transportation and the built environment on the number of civilian vehicles in China. Working paper, University of California at Davis in Davis, California; 2016.

[26] Friedman M, Powell K, Hutwagner L, Graham L, Teague WG. Impact of changes in transportation and commuting behaviors during the 1996 Summer Olympic Games in Atlanta on air quality and childhood asthma. Journal of the American Medical Association. 2001;**285**(7):897–905.

[27] Currie J, Walker R. Traffic congestion and infant health: Evidence from E-ZPass. American Economic Journal: Applied Economics. 2011;**3**:65–90.

[28] Knittel C, Miller DL, Sanders NJ. Caution, drivers! Children present: Traffic, pollution and infant health. Review of Economics and Statistics. 2016;**98**(2):350–366.

[29] Sun C, Zhen S, Wang R. Restricting driving for better traffic and clearer skies: Did it work in Beijing? Transport Policy. 2014;**32**:34–41.

[30] Zhang W, Lin Lawell C-YC, Umanskaya VI. The effects of license plate-based driving restrictions on air quality: Theory and empirical evidence. Journal of Environmental Economics and Management; forthcoming.

[31] Chen Y, Whalley A. Green infrastructure: The effects of urban rail transit on air quality. American Economic Journal: Economic Policy. 2012;**4**(1):58–97.

[32] Lalive R, Luechinger S, Schmutzler A. Does supporting passenger railways reduce road traffic externalities? Discussion Paper No. 9335, Centre for Economic Policy Research; London, UK; 2013.

[33] Harford J. Congestion, pollution, and benefit-to-cost ratios of US public transit systems. Transportation Research, Part D. 2006;**11**(1):45–58.

[34] Holian M, Kahn M. California voting and suburbanization patterns: Implications for transit policy. MTI Report 12-05, Mineta Transportation Institute; San Jose, California; 2013.

[35] Downs A. Still Stuck in Traffic: Coping with Peak-Hour Traffic Congestion. Washington, DC: Brookings; 2004.

Chapter 7

Bus Rapid Transit Systems Road Safety: A Case Study of Mexico City

Jaime Santos-Reyes, Vladimir Ávalos-Bravo and Diego Padilla-Pérez

Additional information is available at the end of the chapter

http://dx.doi.org/10.5772/65583

Abstract

The book chapter presents a statistical analysis of historical data of bus rapid transit (BRT) lines A&B accidents that have occurred in Mexico City from 2005 to 2015. Some of the key conclusions are the following: (a) 484 accidents have occurred when considering both lines A and B. The most critical years have been 2008, 2011 and 2012; the least critical year, on the other hand, has been 2010; (b) overall, the frequency of accident occurrence has been decreasing in both lines; (c) the most critical seasons of the year have been the following: autumn (27.7% in line A) and winter (32% in line B); (d) the frequency of accidents increases when approaching the end of the week (Thursday and Friday) and the frequency of accidents decreases sharply at weekends; (e) 48.28 and 54.47% of accidents have occurred at the three peak (i.e. morning, afternoon, evening/night) in lines A and B, respectively; (f) 64.8% (22/73) of pedestrians have been killed when collided with the BRT buses; and (g) the most critical section of the BRT lane has been identified with 38 (11.87%) accidents and for the case of line A. Future work includes statistical significance tests on the data.

Keywords: accident, bus rapid transit (BRT), statistical analysis, road safety, traffic safety

1. Introduction

The first bus rapid transit (BRT) system was implemented in Curitiba, Brazil, in 1974, and has become a global phenomenon in the twenty-first century [1]. Major new BRT projects have been adopted in many countries worldwide (e.g. Australia, China, India, Indonesia, United States, Iran, Turkey, Europe, etc.); in fact, it is believed that about 168 cities from 39 countries have adopted the system [2, 3]. The success of BRT systems has been attributed to the following advantages [1]: (a) the time of implementation is shorter than for rail-based mode of transport, (b) the implementation costs are a fraction of those for rail-based mass transit system and (c) easy network connectivity, that is, parts of the network can operate on normal streets, it is much cheaper and faster to establish a full network using bus-based mass transit.

However, there is no rigid definition of precisely what constitutes a BRT system. It is thought that the lack of a common definition of BRT has caused confusion in discussions of the technology since its inception [1]. The authors argue that the lack of a common understanding of what constitutes a BRT system has led to branding problems. There is currently no official definition of what constitutes a BRT. Two definitions are as follows:

"A high-quality bus-based transit system that delivers fast, comfortable, and cost-effective urban mobility through the provision of segregated right-of-way infrastructure, rapid and frequent operations, and excellence in marketing and customer service." — ITDP [4].

"An enhanced bus system that operates on bus lanes or other transit ways in order to combine the flexibility of buses with the efficiency of rail....It also utilizes a combination of advanced technologies, infrastructure, and operational investments that provide significantly better service than traditional bus service." — USFTA [5].

On the other hand, research on BRT has been widely reported in the literature on several issues, for example, research from a social perspective [6–11], studies being conducted on the economics of BRT systems [12–14], research has also been conducted from a technical performance perspective [15–20], the environmental impact of the implementation of BRT systems has also been addressed by several authors [13, 16, 21], and some aspects on road safety have been reported in the literature, for example, see [6, 15].

However, there is no evidence of studies being conducted explicitly on accidents associated with BRT systems. This may be relevant in understanding, for example, deficiencies in the current rod safety associated with BRT systems. Further, it may help to better understand urban mobility. The book chapter gives an account of the ongoing research project concerning a statistical analysis of historical accident data analysis associated with BRT systems in Mexico City, in particular, those accidents that have occurred in BRT lines A&B, and for the time between 2005 and 2015.

2. BRT systems worldwide and in Mexico City

2.1. BRT systems worldwide

As mentioned in the 'Introduction' section, the BRT system emerged as an option to the transport problem in some developing cities in Latin America. However, BRT systems have not only become popular in Latin America but also become a popular means of transport worldwide, due its value, serviceability, accessibility, flexibility relative and network coverage [1, 2, 19]. Further, many studies have shown that BRT systems can be a cost-effective way to provide a service of high-performance transport [22, 23].

For example, the system has been expanded in such a way that currently BRT systems are in operation in 206 cities spread over five continents (**Figure 1**). Further, it is believed that the BRT systems worldwide transport 33.3 million passengers per day [2]. Furthermore, the world's leading regions are Latin America and Asia, covering 108 cities. **Figure 1** shows the countries that have invested heavily in BRT transport systems, for example, Brazil (34 cities, 847 km of length), Mexico (11 cities, 340 km) and Colombia (six cities, 205 km).

2.2. BRT systems in Mexico City

The Metropolitan area of Mexico City has been regarded as the third largest agglomeration in the world. It is believed that the population is about 21.4 million people; further, the population density is about 6000 inhabitants/km^2 [24]; it is thought that about 600 new motor cars are being registered each day. The Metro underground mass transit system served for a long time as the dominating mode of transportation with approximately 250 km of length and 4.5 million trips per day [24]. Further, the traditional transport systems were typically composed of overcrowded and slow private bus and minibus lines. Furthermore, the buses were old, poorly maintained and highly polluted. These transport shortcomings prompt the capital City's decision-makers to solve the problem with investments in road infrastructure; initially, they built additional motorways; however, it was clear that those measures induced more traffic congestion and the project only served a small percentage of the city's inhabitants. It is thought for this reason; the idea of implementing BRT systems came up as the best solution for solving congestion and air pollution problems [24] (see **Figure 1** and **Table 1**).

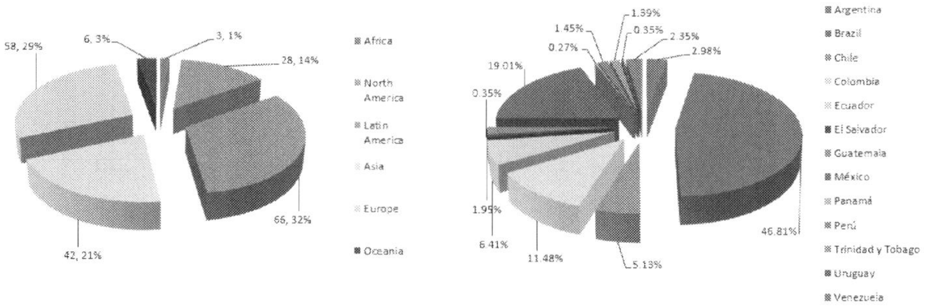

Figure 1. Number of cities worldwide that have adopted BRT transport systems.

Characteristics	Line A	Line B	Line C	Line D	Line E	Total
Length	30 km	20 km	17 km	28 km	10 km	**105 km**
Starting operations	June 2005	December 2009	February 2011	April 2012	November 2013	–
No. of stations	44	34	29	32	16	**155**
Passengers/day	480,000	180,000	140,000	65,000	55,000	**920,000**

Table 1. Some characteristics of the four BRT lines in Mexico City [25].

In 2002, the Mexico City government planned a BRT corridor running across the centre of the capital city. In 2005, the first corridor of the BRT line (i.e. BRT line A) started operations along one of the key avenues with 30 km of length (**Figure 2** and **Table 1**). It is thought along this avenue, the BRT system has improved mobility by 50%. The success of line A prompted the opening of line B; in 2008, along another of the key avenues of the city, this was followed by the opening of line C, in 2011 [24]; a year later, line D started its operations (April 2012). Further extensions are already implemented and others are planned and in construction. **Figure 2** shows a map of the four lines and **Table 1** shows some of the key features of the two BRT lines being considered in the present analysis.

Figure 2. Four BRT lines in Mexico City.

3. Analysis and results

A BRT system database-related accident in Mexico City has been built. The database covers accident data since the opening of the first BRT system in 2005 until 2015. The historical data have been collected through the following sources: (a) data from the organization running the system [25], (b) incident/accident reports from the police and (b) data being collected from the mass media reports when accidents occur (i.e. TV, newspapers, online news, etc.).

Overall, a study associated with traffic accidents consists of analysing different variables intended to provide insight into their behaviour, location of occurrence (i.e. urban and sub-urban), the time of occurrence, date, class and type of accident, type of vehicle involved, the causes of the accident, driver details and class of victims (i.e. fatal and non-fatal). In this particular case, all the accidents are related to urban location. In what follows, some of the variables considered in the analysis are listed below [26]:

a. Accident occurrence time (year, month, weekday, hour).

b. Type of traffic accident (fatal and not fatal).

c. Vehicle involved in the collision with a BRT (motorcar, passenger van, cargo van, cargo truck, minibus, motorbike and bicycle).

d. Type of collision (between the BRT unit and motorcar, pedestrian, motorbike and cyclist).

e. Victims (driver, passenger, pedestrian and cyclist).

f. Age and gender of the victims.

g. Location of the accidents.

The variable associated with the immediate cause of the accidents has not been considered in the analysis presented here. It may be argued that an accident has multiple causes and needs to be fully analysed, see, for example, [27]. In the subsequent subsections, the main results associated with these variables are presented and the most relevant results are discussed in Section 4.

3.1. BRT accidents per year

As mentioned in Section 2.2, the BRT line A started operations in 25, whereas line B in 28 (**Table 1**). It has been found that 484 accidents have occurred in lines A&B combined. That is, 350 in line A (**Figure 3a**) and 134 in line B (**Figure 3b**). When considering line A separately, there has been a mean of 31.8 accidents per year. Further, **Figure 3a** shows a tendency of decreasing the frequency of accidents in this particular line. On the other hand, line B registered a mean of 16.7 accidents per year.

From **Figure 3**, it also can be observed that in both lines A and B, the frequency of the occurrence of accidents increased in the subsequent years once they have initiated operations. For example, the frequency of occurrence of accidents in line A reached the highest, 4 years after initial operations (i.e. 2008) with a total of 66 (18.85%; 66/350). Similar results have been seen for the case of the BRT line B in 2009 (23.8%; 32/134). Interestingly, both lines show a similar increasing trend in accident occurrence in the most recent years, that is, 2014–2015 (line A) and 2013–2015 (line B). Finally, it also can be seen that the year 2010 has been the least critical year for both lines (i.e. 10 accidents registered in line A and 8 in line B).

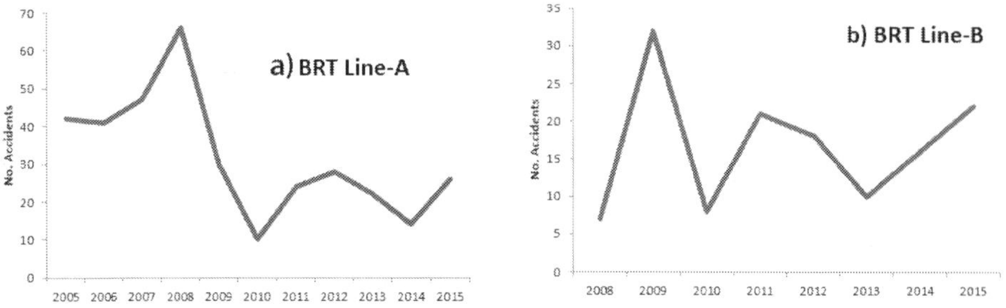

Figure 3. Accidents per year: (a). BRT Line-A; (b). BRT Line-B.

3.2. Accidents per month

Figure 4 shows the frequency of accident occurrence per month for the two BRT lines considered in the present analysis. Overall, it can be seen that line A has an increasing tendency of accident occurrence towards the end of the year; April has been the lowest accident occurrence with 5.1% (18/350) and November registered the highest number of accidents with 12% (42/350). Line B, on the other hand, shows a decreasing tendency of accident occurrence towards the end of the year. December registered the highest number with 12.68% (17/134) and July registered the lowest with 4.47% (6/134).

In an attempt to understand the trend of accident occurrence, for example, in line A (i.e. the oldest of the two lines being considered in the analysis), the data have been plotted by month and year. The results are shown in **Figure 5**. It can be seen that the highest frequency of occurrence has been the following months: August (2005) and November (2008) with 11 and 12 accidents, respectively. As mentioned in the previous sections, 2010 has been a year with the lowest number of accidents, that is, only 10 events have been registered. On the other hand, the year 2008 registered the highest number of accidents (18.85%; 66/350). In fact, zero accidents have been registered in 2010 and for the following months: January, June, August, September and October.

Another aspect that should be highlighted by observing at **Figure 5** is the fact that November has been the month with the highest frequency of accidents 12% (42/350) when considering

http://dx.doi.org/10.5772/65583

the 11 years altogether, that is, 2005–2015. On the other hand, it has been found that April has been the month with the lowest number of accidents for the same period of time, that is, 18 (5.1%; 18/302).

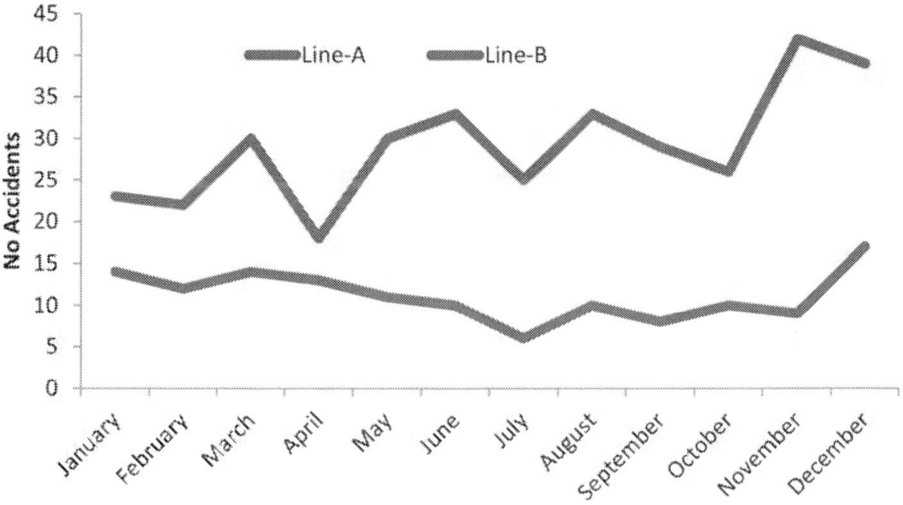

Figure 4. BRT accidents per month.

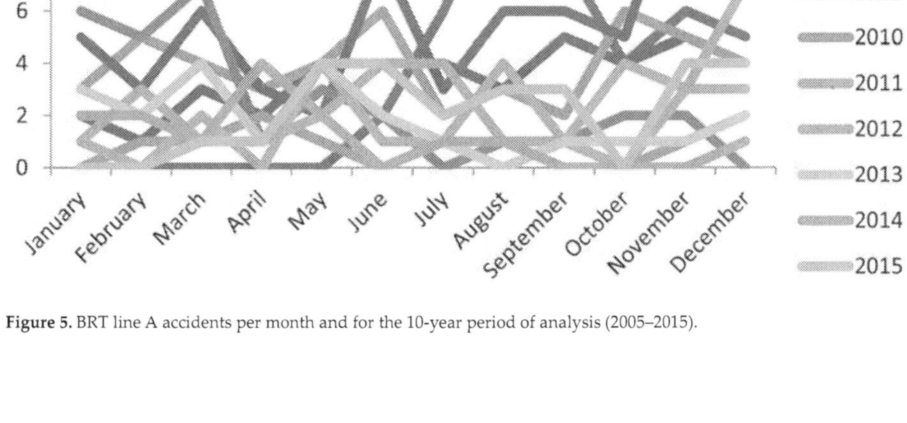

Figure 5. BRT line A accidents per month and for the 10-year period of analysis (2005–2015).

Finally, it was decided to investigate what were the most critical seasons of the year regarding accident occurrence. The results are shown in **Figure 6**. It can be seen that 27.7% (97/350) of accidents occurred during autumn for the case of line A. On the other hand, winter has been the critical season of the 8-year period for the case of line B (32%; 43/134). It is interesting to see how the pattern of accident occurrence is opposing when considering the seasons of the year for both lines. This may require further research to understand what are the 'external' factors that have an influence on this.

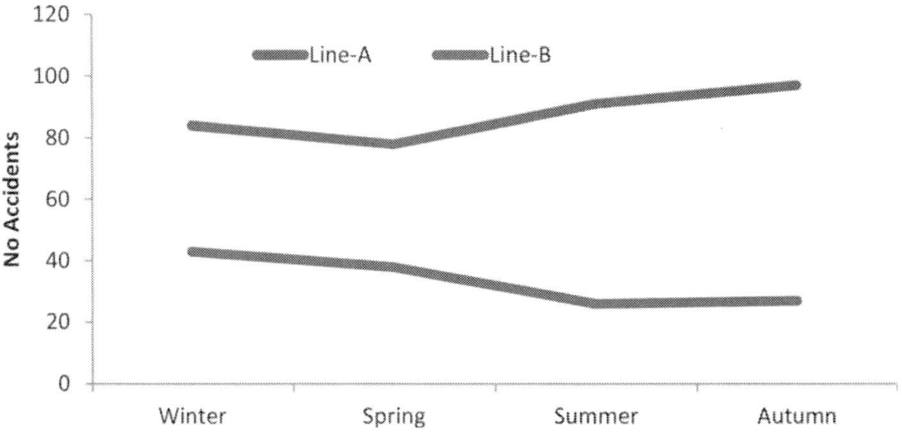

Figure 6. BRT accidents per season of the year.

3.3. Accidents per weekdays

In an attempt to identify some trends of accident occurrence during the days of the week, an analysis of the data was conducted to address this very issue. **Figure 7a** shows the results of the analysis. Overall, it can be seen that the frequency of accidents tends to increase when reaching Friday for the case of line A (23.7%; 83/350).

This is followed by Wednesday and Thursday with 70 (20%; 70/350) and 49 (14%; 49/350) accidents, respectively. The most critical weekdays for the case of line B, on the other hand, have been Thursday (22.3%; 30/134), Tuesday (20.8%; 28/134) and Wednesday (19.4%; 26/134), as shown in **Figure 7b**. Further, the data show that in both lines there is a sharp decreasing tendency of accidents at the weekends (**Figure 7**).

3.4. Accidents per hour

As with the previous sections, a detailed analysis of the data being collected has been conducted for the distribution of accidents per hour. The range of the BRT operational time has been considered from 04:00 h to Midnight. **Figure 8** and **Tables 2** and **3** show the results of

http://dx.doi.org/10.5772/65583

the analysis. The distribution of accidents that have occurred in line A per hour and for the period between June 2005 and 31 December 2015 can be seen. Overall, the distribution of accident occurrence for this particular BRT line can be explained on the basis the following three ranges of time (**Figure 8**): (a) 04:00–13:59 h, (b) 14:00–19:59 h and (c) 20:00–24:00 h. In the time range (a), an increase in the occurrence of accidents is observed, reaching a maximum of 38 (10.85%; 38/350) and subsequently reduced to 36 accidents. The figure shows the most critical period in accident occurrence in the case (b). That is, 171 accidents have been reported, representing 48.8% of the total that occurred in this line. From **Figure 8**, it can also be seen that between 16:00 and 17:59 h, there was the peak of accidents with 63 (18%; 63/350). Finally, the time range (c), a very sharp drop of accident occurrence, can be seen from **Figure 8**.

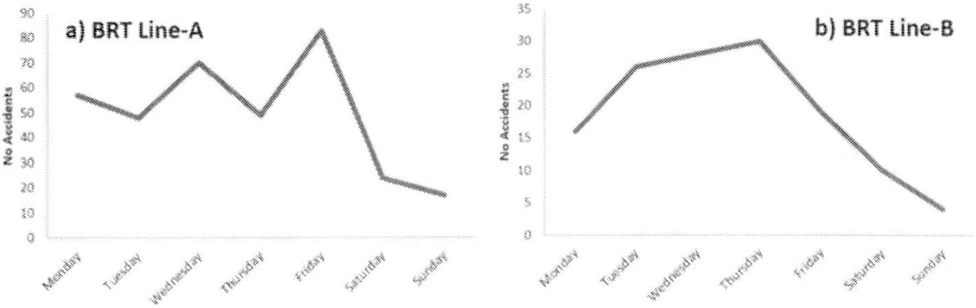

Figure 7. Distribution of accidents per day: (a). BRT Line-A; (b). BRT Line-B.

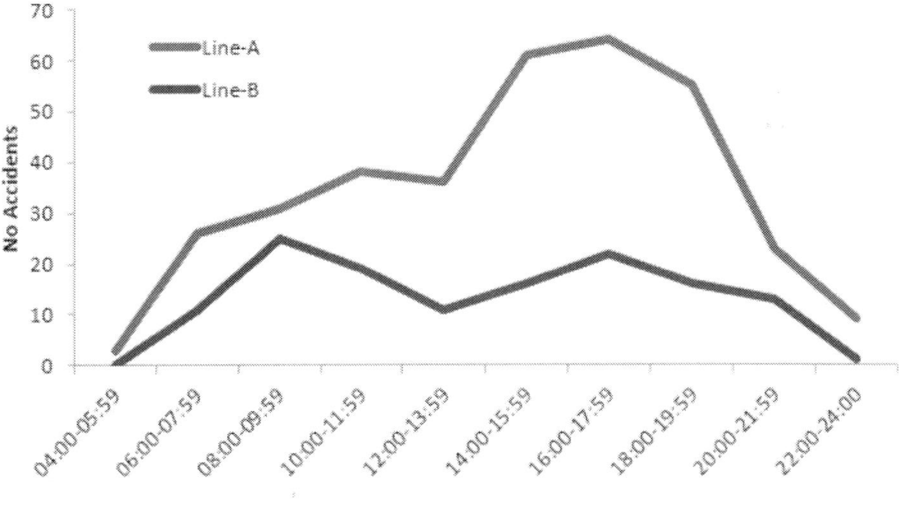

Figure 8. Accidents per hour.

Time period	2005	2006	2007	2008	2009	2010	2011	2012	2013	2014	2015	Total
04:00–05:59	0	0	0	0	1	0	0	1	0	0	1	**3**
06:00–07:59	2	2	2	3	5	1	2	5	1	1	2	**26**
08:00–09:59	4	3	5	7	6	2	0	1	2	0	2	**32**
10:00–11:59	7	3	3	7	3	2	3	3	3	3	2	**39**
12:00–13:59	3	2	3	6	4	2	4	3	3	3	3	**36**
14:00–15:59	7	7	10	12	1	2	3	6	8	2	3	**61**
16:00–17:59	6	9	6	14	7	1	5	5	3	4	4	**64**
18:00–19:59	9	10	10	16	1	0	2	1	2	0	4	**55**
20:00–21:59	4	2	6	1	1	0	4	3	0	0	2	**23**
22:00–Midnight	0	3	2	0	1	0	1	0	0	1	2	**1**
Unknown	0	0	0	0	0	0	0	0	0	0	1	**1**
Total	**42**	**41**	**47**	**66**	**30**	**10**	**24**	**28**	**22**	**14**	**26**	**350**

Table 2. Accidents per hour and year in line A.

The results of the distribution of accidents per hour during the 8-year period between 2008 and 2015, for the case of line B, are also shown in **Figure 8** and **Table 3**. The highest number of accidents occurred during the following ranges of time: (a) 08:00–09:59 h with 25 (18.65%; 25/134) accidents and (b) 20 (14.92%; 20/134) accidents occurred in the time range of 16:00–17:59 h. The lowest number of accidents reported during the afternoon was 11 (8.2%), which occurred in the time range from 12:00 to 13:59 h. Finally, the range of time is from 20:00 to 24:00 h, a very sharp decline in the occurrence of accidents, that is, 14 (10.44%).

Time period	2008	2009	2010	2011	2012	2013	2014	2015	Total
04:00–05:59	0	0	0	0	0	0	0	0	**0**
06:00–07:59	1	1	2	2	1	3	0	1	**11**
08:00–09:59	2	7	2	4	2	2	3	2	**25**
10:00–11:59	2	9	1	1	0	0	1	3	**19**
12:00–13:59	0	2	0	0	5	1	1	2	**11**
14:00–15:59	0	4	1	3	2	2	3	1	**16**
16:00–17:59	1	4	1	4	4	1	3	3	**22**
18:00–19:59	1	3	1	4	1	1	2	3	**16**
20:00–21:59	0	2	0	3	2	0	3	2	**13**
22:00–Midnight	0	0	0	0	1	0	0	0	**1**
Unknown	0	0	0	0	0	0	0	0	**0**
Total	**7**	**32**	**8**	**21**	**18**	**10**	**16**	**22**	**134**

Table 3. Accidents per hour and year in line B.

3.5. Accidents on peak hours

An analysis of the BRT line users data was conducted; it has been found that for line A, there are three peak hours during the operational day: morning (06:00–08:00 h), afternoon (13:00–15:00 h) and at evening/night (17:00–20:00 h). Overall, 48.28% (169/350) of accidents occurred during the three ranges of peak hours. The results also show that the evening/night peak hours are where 14.23% of accidents have occurred.

As with the case of line A, the same analysis was performed for line B. The results show that 54.47% (73/134) of all the line accidents occurred within three ranges of peak hours in this line. Similarly, 25% of the accidents occurred in the evening/night peak hours.

Effectively, these accidents have affected traffic congestion and consequently the urban mobility. See Section 4.4 for the discussion about this.

3.6. Victims of the accidents

Figure 9 shows the results of the victims as a result of the collisions with BRT buses. Given the lack of data from 2005 to 2007, it should be pointed out that the results shown in **Figure 9** are data available only from 2008 to 2014. The results show that a high percentage of collisions have been those associated with BRT units and pedestrians with 64.8% (64/73). Those related to collisions with cyclists with 8.1% followed this; finally, 1.1% are related to collisions with motorbike users. **Figure 9** also shows that the highest number of collisions with pedestrians occurred in 2012 with a total of 15 (20.5%; 15/73).

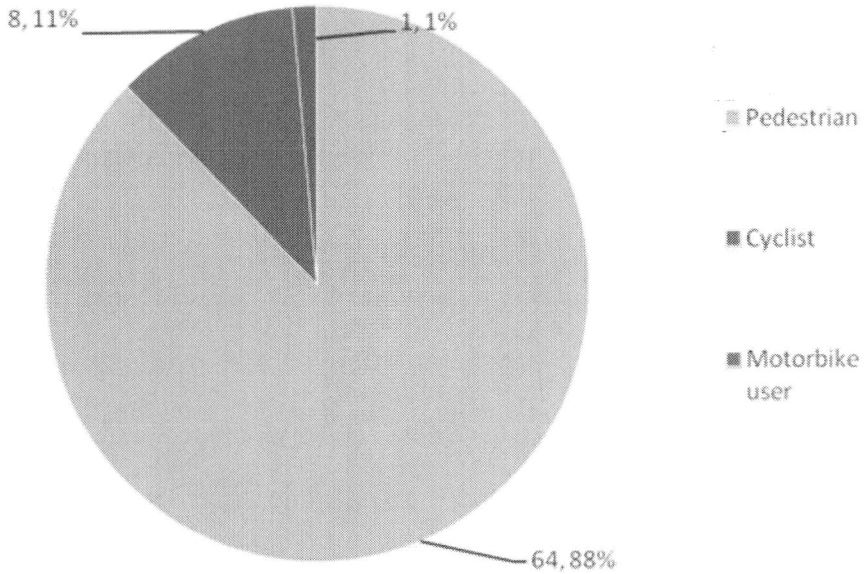

Figure 9. Victims of the accidents.

3.7. Fatal versus non-fatal

This subsection presents the results of the degree of injuries as results of the collision between the pedestrians and BRT units (**Table 4**). As mentioned in the previous sections, the results presented here are limited to 73 registered data associated with the consequences of such collisions (2008–2014). There have been 22 (30.1%; 22/73) fatal incidents and 51 (69.9%; 51/73) non-fatal accidents as a consequence of the collisions. **Table 4** shows the results associated with pedestrians being killed by the collisions; it can be seen that 45.5% have been killed in BRT line A with the highest percentage of those being male.

Line	Gender	10–14yo	15–19yo	20–24yo	25–29yo	30–34yo	35–39yo	40–44yo	45–54yo	55–64yo	Over 65yo	Total
Line A	MF	00	10	10	00	00	00	00	20	11	00	**51**
Line B	MF	00	10	00	11	10	01	00	00	11	10	**53**
	Total	**0**	**2**	**3**	**2**	**1**	**1**	**0**	**3**	**4**	**1**	**22**

F = Female, M = Male, yo = years old.

Table 4. Fatal pedestrian collisions.

The results also show that the highest frequency of fatalities occurred to those between the ages of the following ranges: 20–24 and 45–54 with three being killed in each category.

When considering individual lines, the highest frequency of occurrence is associated with line A with 12 (23.5%). The results also show that the most vulnerable population to collisions are young pedestrians in the range between 15 and 19 years old.

3.8. Location of the accidents on the BRT corridor

This subsection presents the results associated with the identification of the BRT lanes where the accidents occurred. For illustrative purposes, only the results regarding line A are shown here (**Figures 10** and **11**). The top five BRT lane sections with the highest number of accident occurrence were the following (**Figure 11**):

1. Lane 15 between stations 14 and 15 (L15 (S14-S15)) with 38 (11.87%).
2. Lanes 12 and 13 with 22 (6.87%) each.
3. Lane 22 with 18 (5.62%).
4. Lane 17 with 17 (5.31%).
5. Lanes 36 and 37 with 15 (4.687%) each.

http://dx.doi.org/10.5772/65583

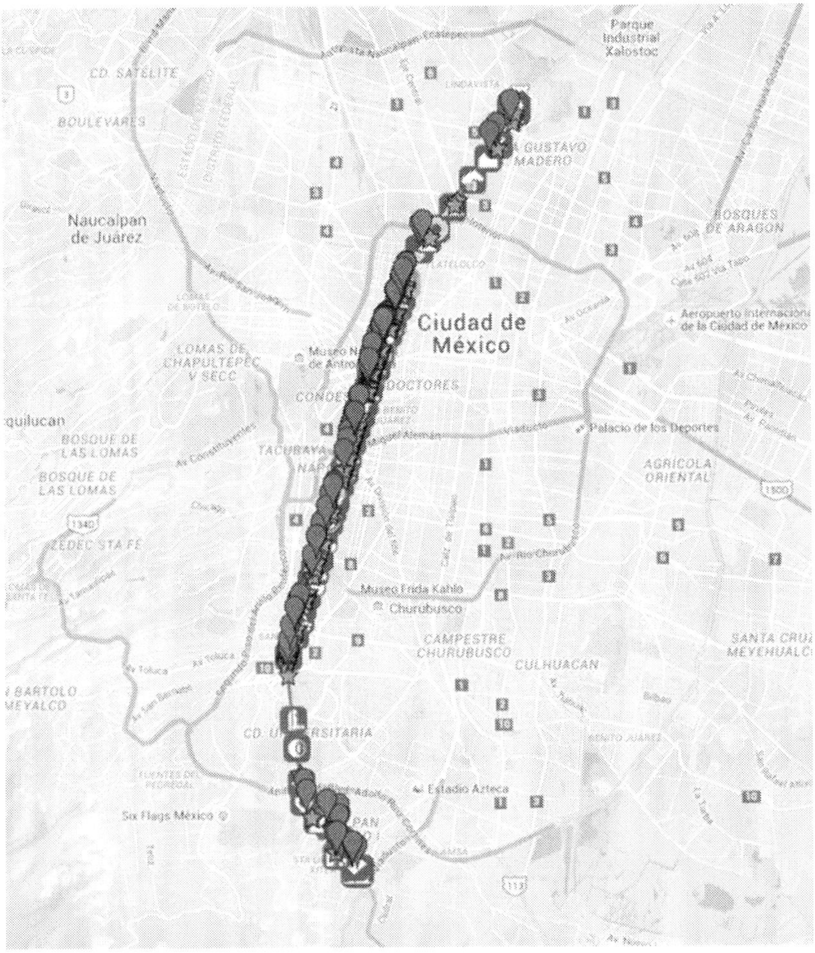

Figure 10. Location of the accidents for the case of the BRT line A.

4. Discussion and conclusion

One of the biggest challenges facing megacities, such as Mexico City, lies in lagging infrastructure. That is, these cities continue to add population, without the infrastructure paralleled the growth [28]. An example of a lack of infrastructure is that related to transportation [28]. Further, in the ultra-dense environment of developing country megacities, traffic congestion

Figure 11. Distribution of the accidents that occurred in the lanes (L) between the stations (S) of the BRT line A.

is also worsening and effectively affecting the urban mobility [28, 29]. In recent years, Mexico City has implemented a number of policies aiming at improving the transport infrastructure by adopting, for example, bus rapid transit (i.e. Metrobus) systems.

Given the above, it becomes essential to prevent accidents associated with BRT systems in order to improve, among other things, road safety associated with BRT systems and therefore the urban mobility of the inhabitants of the capital city (**Figure 12**). In an attempt to understand BRT-related accidents, the authors have built a BRT accident database for the case of Mexico City. Some results of a statistical analysis of these have been presented here. The aim has been to understand, *inter alia*, the general pattern of the frequency of accidents occurrence during the day, weekdays, months, and year and to locate the BRT lanes of major occurrence. However, the variable associated with the cause of these accidents has not been considered here (however, it should be mentioned that the immediate cause of most of the accidents has been that motor cars at regular lanes crossed over BRT lanes). That is, it may be argued that BRT-related accidents should be fully analysed to understand the causal factors leading to their occurrence; this may be the only way to gain a better understanding of the multicausal factors and prevent future recurrence [27]. This may be regarded as a limitation of the present analysis. It also should be mentioned that given the lack of data some of the variables such as pedestrians being killed during the collisions are reported from 2008 to 2015 only (Sections 3.6 and 3.7).

Figure 12. Example of a BRT accident in the city and affecting urban mobility [33].

Overall, it has been found that there have been 484 accidents when considering both lines A and B and the most critical years have been 2008, 2011 and 2012; the least critical year, on the other hand, has been 2010 (Section 3.1). The latter is quite surprising given the fact that by this year there were the two lines operating (lines A and B). However, at this stage we are unable to explain as to why 2010 registered only 18 accidents. The results also show a general tendency of decreasing the frequency of accident occurrence for the lines considered in the analysis. This may be explained given the fact that the road conditions, among other things, have been improved in recent years. For example, a study associated with an assessment of road conditions found that there were many deficiencies associated with road infrastructure endangering public safety [30]. For example, the report stresses that there had been only one traffic light to control the pedestrians crossing in several wide intersections. Further, it is believed that pedestrians are expected to take 11 s during the crossing; no pedestrian traffic lights and sidewalks in poor condition and with no ramps for the disabled were found in the study. The report also found that unfortunately heavy good vehicles (HGVs) still are allowed to circulate in avenues where Metrobus circulates in counterflow and carrying up to 150 passengers on board. Effectively, this (HGV) represents an additional risk factor for accidents. The human factors component is also crucial in road accident occurrence [31]. The CENAPRA [30] report found that pedestrians carelessly cross (the street) endangering their safety; further, the results of BRT accident analysis have been found that human error has been a contributing factor for accident occurrence; motor car drivers (HGV and BRT drivers) very often ignored the red lights [27].

The results have also highlighted that when considering the accident occurrence by month, the most critical seasons of the year have been during autumn (line A: 27.7%; 97/350), and winter (line B: 18%; 18/134). Again, at this stage of the ongoing research project, we are unable to comment on the reasons for this; however, we can say that by conducting a comprehensive accident analysis we may be able to find the causal factors of the accidents and shed some light on the reasons for this [27]. It also has been found that the frequency of accidents at weekends decreases sharply for both lines (Section 3.3). This was expected given the fact that most of the commuters avoid going to the city centre and prefer staying home and this is in line with the number of riders reported at the weekends.

However, what is clear is that when accidents occur, megacities such as Mexico city usually cause traffic congestion which in turn causes slower speeds, longer travel times and increased vehicular queuing, that is, when traffic demand is great enough that the interaction between vehicles slows the speed of the traffic stream, this results in congestion (**Figure 12**). Another of the findings of the study is that accidents occur throughout the whole BRT operational hours (i.e. 06:00 till midnight). In other words, accidents have occurred in peak hours, for example, it has been found that 48.23% (169/350) of accidents have occurred at the three peak hours in line A; 14.23% occurred at evening/night peak hours. Similarly, 25% occurred for the same peak hours in line B (Section 3.5). This raises the following question: do BRT accidents affect mobility in megacities such as Mexico city? Vermeiren [32] argue that urban growth decreases individual mobility and argue, "An individual is considered highly mobile when he or she is able to easily and comfortably reach his or her destination(s) in space and time." Effectively, this is dependent, among other things, on the city's transportation network free of accidents. The following examples illustrate what happens when accidents associated with BRT occur in Mexico City [33]:

"Although there were no injured the (BRT) service was interrupted as passengers had to descend and wait another (BRT) unit. This accident further complicated traffic on the 4 South Av which was particularly affected by the increase in vehicle load as a result of the day back to school."

"A (BRT) collision occurred at Insurgentes Av..., causing traffic chaos...The police arrived on the scene to expedite the affected traffic in the area."

Further, Kotkin [28] argue that traffic congestion has a number of negative effects, for example, (a) motorists and passengers lost time. As a non-productive activity for the affected people, congestion reduces the city's economic health; (b) delays, which may result in late arrival for employment, meetings and education, resulting in lost business, disciplinary action and other personal losses; (c) wasted fuel increasing air pollution and CO_2 emissions owing to increased idling, acceleration and braking; (d) affectation of emergency services, for example, blocked traffic may interfere with the passage of emergency vehicles travelling to their destinations where they are urgently needed; (e) higher probability of collisions due to tight spacing and constant stopping and going.

In summary, the results presented here have shown a number of trends that may help to better plan to prevent congestion and improve, among other things, urban mobility in the Capital City. Some of the key conclusions are the following:

a. Four hundred and eighty-four accidents have occurred when considering both lines A and B. The most critical years have been 2008, 2011 and 2012; the least critical year, on the other hand, has been 2010.

b. Overall, the frequency of accident occurrence has been decreasing in both lines.

c. The most critical seasons of the year have been the following: autumn (27.7% in line A) and winter (32% in line B).

d. The frequency of accidents increases when approaching the end of the week (Thursday and Friday) and the frequency of accidents decreases sharply at weekends.

e. 48.28 and 54.47% of accidents have occurred at the three peaks (i.e. morning, afternoon and evening/night) in lines A and B, respectively).

f. 64.8% (22/73) of pedestrians have been killed when collided with the BRT buses.

g. The most critical section of the BRT lane has been identified with 38 (11.87%) accidents and for the case of line A.

Acknowledgements

The project was funded under the following grant: SIP-IPN: No. 2016149.

Author details

Jaime Santos-Reyes*, Vladimir Ávalos-Bravo and Diego Padilla-Pérez

*Address all correspondence to: jrsantosr@hotmail.com

Grupo de Investigación SARACS, ESIME, ZAC, Instituto Politécnico Nacional, México City, México

References

[1] Weinstock, A., Hook, W., Replogle, M., Cruz, R. Recapturing Global Leadership in Bus Rapid Transit-A Survey of Select U.S. Cities. USA: Institute for Transportation & Development Policy; 2011.

[2] Global BRT Data. Produced by bus rapid transit across latitudes and cultures and EMBARQ, in partnership with IEA and SIBRT [Internet]. 2014. Available from: http://brtdata.org. [Accessed: 1 June 2016].

[3] Hidalgo, D. Bus rapid transit: worldwide history, key systems and future directions. Encyclopedia of sustainability science and technology [Internet]. 2011. Available from: http://www.springerreference.com/docs/html/chapterdbid/308766.html. [Accessed: 1 June 2016].

[4] ITDP (Institute for Transportation and Development Policy). Bus Rapid Transit Planning Guide. 3rd ed. USA: ITDP; 2007.

[5] USFTA (U.S. Federal Transit Administration). Boston Silver Line Waterfront Bus Rapid Transit (BRT) 2007 Project evaluation [Internet]. 2007. Available from: http://www.docstoc.com/docs/857934/Boston-Silver-Line-Waterfront-Evaluation. USA [Accessed: March 23, 2011].

[6] Yazici, M., Levinson, H., Ilicalli, M., Camkasen, N., Kamga, C. A bus rapid transit line case study: Istanbul's Metrobus System. Journal of Public Transportation. 2013;**16**(1):153–177.

[7] Delmelle, E.C., Casas, I. Evaluating the spatial equity of bus rapid transit-based accessibility patterns in a developing country: The case of Cali, Colombia. Transport Policy. 2012;**20**:36–46. DOI: http://dx.doi.org/10.1016/j.tranpol.2011.12.001

[8] Lin, Z., Wu, J. Summary of the application effect of bus rapid transit at Beijing south-centre corridor of China. Journal of Transportation Systems Engineering & Information Technology. 2007;**7**(4):137–142. DOI: http://dx.doi.org/10.1016/S1570-6672(07)60034-8

[9] Mejía-Dugand, S., Hjelm, O., Baas, L., Alberto Ríos, R. Lessons from the spread of bus rapid transit in Latin America. Journal of Cleaner Production. 2014;**50**:82–90. DOI: http://dx.doi.org/10.1016/j.jclepro.2012.11.028

[10] Rizvi, A., Sclar, E. Implementing bus rapid transit: A tale of two Indian cities. Research in Transportation Economics. 2014;**48**:194–204. DOI: http://dx.doi.org/10.1016/j.retrec.2014.09.043

[11] Munoz, J.C., Batarce, M., Hidalgo, D. Transantiago, five years after its launch. Research in Transportation Economics. 2014;**48**:184–193. DOI: http://dx.doi.org/10.1016/j.retrec.2014.09.041

[12] Tirachini, A. The economics and engineering of bus stops: Spacing, design and congestion. Transportation Research-Part A. 2014;**59**:37–57. DOI: http://dx.doi.org/10.1016/j.tra.2013.10.010

[13] Cervero, R., Kang, C.D. Bus rapid transit impacts on land uses and land values in Seoul, Korea. Transport Policy. 2011;**18**:102–116. DOI: http://dx.doi.org/10.1016/j.tranpol.2010.06.005

[14] Hensher, D.A., Golob, T.F. Bus rapid transit systems: A comparative assessment. Transportation. 2008;**35**:501–518. DOI: http://dx.doi.org/10.1007/s11116-008-9163

[15] Hidalgo, D., Lleras, G., Hernández, E. Methodology for calculating passenger capacity in bus rapid transit systems: Application to the Transmilenio system in Bogotá, Colombia. Research in Transportation Economics. 2012;**39**(1):139–142. DOI: http://dx.doi.org/10.1016/j.retrec.2012.06.006

[16] Nugoro, S.B., Fujiwara, A., Zhang, J. The influence of BRT on the ambient PM10 concentration at roadside sites of Trans Jakarta Corridors. Procedia Environmental Sciences. 2010;**2**:914–924. DOI: http://dx.doi.org/10.1016/j.proenv.2010.10.103

[17] Currie, G., Delbosc, A. Assessing bus rapid transit system performance in Australasia. Research in Transportation Economics. 2014;**48**:142–151. DOI: http://dx.doi.org/10.1016/j.retrec.2014.09.012

[18] Clifton, G.T., Mulley, C., Hensher, D.A. Bus rapid transit versus heavy rail in suburban Sydney—Comparing successive iterations of a proposed heavy rail line project to the pre-existing BRT network. Research in Transportation Economics. 2014;**48**:126–141. DOI: http://dx.doi.org/10.1016/j.retrec.2014.09.010

[19] Hensher, D.A., Li, Z., Mulley, C. Drivers of bus rapid transit systems e influences on patronage and service frequency. Research in Transportation Economics. 2014;**48**:159–165. DOI: http://dx.doi.org/10.1016/j.retrec.2014.09.038

[20] Tao, S., Corcoran, J., Mateo-Babiano, I., Rohde, D. Exploring bus rapid transit passenger travel behaviour using big data. Applied Geography. 2015;**53**:90–104. DOI: http://dx.doi.org/10.1016/j.apgeog.2014.06.008

[21] Wöhrnschimmel, H., Zuk, M., Martínez-Villa, G., Cerón, J., Cárdenas, B., Rojas-Bracho, L., Fernández-Bremauntz, A. The impact of a bus rapid transit system on commuter's exposure to Benzene, CO, PM2.5 and PM10 in Mexico City. Atmospheric Environment. 2008;**42**:8194–8203. DOI: http://dx.doi.org/10.1016/j.atmosenv.2008.07.062

[22] Levinson, H., Zimmerman, S., Clinger, J., Gast, J., Rutherford, S., Bruhn, E. Implementation guidelines. Transit Cooperative Research Programme Report 90. In: Transportation Research Board, editor. Bus Rapid Transit (Vol. 2). Washington, D.C.: National Academies; 2003.

[23] Wright, L., Hook, W. Bus Rapid Transit Planning Guide. New York, NY, USA: Institute for Transportation & Development Policy; 2007.

[24] MB (Mercedes-Benz). BRT in Mexico City, Mexico: Metrobus linea 3 corridor – An example of a successful BRT solution in one of the biggest cities in the world. Mexico: MB (Mercedes-Benz); 2013.

[25] Metrobus. Fichas técnicas [Internet]. 2016. Available from: http://www.metrobus.cdmx.gob.mx/fichas.html [Accessed: 5 June 2016]

[26] INISG (National Institute of Statistics and Geography). Methodological summary of road traffic accidents statistics in urban and suburban areas. Mexico: NISG; 2009.

[27] Santos-Reyes, J., Ávalos-Bravo, V. A preliminary analysis of two bus rapid transit accidents of Mexico City's transport system. Procedia Engineering. 2014;**84**:624–633. DOI: http://dx.doi.org/10.1016/j.proeng.2014.10.479

[28] Kotkin, J., Cox, W., Modarres, A., Renn, A.M. The Problem with Megacities. USA: Chapman University Press; 2003.

[29] Heinrichs, D., Krellenberg, K., Hansjürgens, B., Martínez, F. Risk Habitat Megacity. Springer; 2012. 366 p. DOI:10.1007/978-3-642-11544-8.

[30] NCIP (National Center for Accident Prevention). Road safety audit in Metrobus line 2. Mexico: Secretary for Prevention and Health Promotion; 2009.

[31] AASHTO (American Association of State Highway and Transportation Officials). A Policy on Geometric Design of Highways and Streets. USA: AASHTO; 2001.

[32] Vermeiren, K., Verachter, E., Kasaija, P., Loopmans, M., Poesen, J., Rompaey, A.V. Who could benefit from a bus rapid transit system in cities from developing countries? A case study from Kampala, Uganda. Journal of Transport Geography. 2015;**47**:13–22

[33] El Universal. Carambola de transporte público en el D.F. Available from: http://fotos.eluniversaldf.mx/coleccion/muestra_fotogaleria.htmal?idgal=13339 [Accessed: 5 June 2016].

Chapter 8

Procedure for the Award of Contracts and Contracting in Public Passenger Transport

Miloš Poliak

Additional information is available at the end of the chapter

http://dx.doi.org/10.5772/64820

Abstract

This chapter describes a procurement process for providing transport serviceability by public passenger transport. The objective of the chapter is to present individual steps for procurement of public transport services. These steps consist of identification of objectives, definition of requirements for transport serviceability, risk allocation between contractual parties, drafting a public service contract and a process of selecting a service operator. Special attention is paid to the risks and their influence on contracting parties. The chapter also characterises procedure for the direct award of a public service contract, that is without competitive tendering. The author tries to define the impact of the direct award of contracts on the scope of services provided.

Keywords: procedure, contract, public, transport, risk

1. Introduction

Properly concluded contractual relationship allows the creation of a strong partnership through which the authorities can pursue their policy objectives. Such a partnership should prevent from neglecting fulfilment of the tasks or abusing position of one from parties. The key factor for providing public transport services is an adequate regulatory framework and contracting conditions that should be set to support the competitive behaviour of bidders—service providers.

The regulatory framework consists of three levels [1]:

- *Strategic* (setting basic objectives to be achieved: transport policy, public budgets, intermodality, etc.).

- *Tactical* (emphasis mainly on design of services and fares, requirements for staff, vehicles, additional services).
- *Operational* (ensuring the service provision in the market according to objectives: sale activities, information for the public, deployment of vehicles, maintenance).

In the process of preparing and contracting, the attention must be paid to aspects such as [2]:

- setting the rights and obligations of the parties,
- addressing the issue of infrastructure ownership and ownership of service provider (transport means and depots),
- risk allocation between the parties,
- responsibility for planning and design of services to be provided (freedom which be given to the operator in designing and making changes in services),
- scope of contract,
- structure of payment (it should represent a balance between profit and reward for realised performance),
- method of monitoring and controlling a fulfilment of public service obligations.

It should be noted that a proper adjustment of these aspects by authorities helps to operators improve efficiency of service provision and reduce costs.

1.1. Addressing the issue of ownership

The provision of public transport services requires on the one hand the availability of particular assets (such as infrastructures and vehicles), and on the other hand the management of those assets in combination with personnel.

There are several possibilities of ownership [3]:

- public ownership,
- *mixed ownership* majority private partner of more than 50% or minority private partner of less than 50%,
- private ownership.

Infrastructure ownership and ownership of service provider can be separated according to following ways (and the ownership can be organised differently for each part):

- vertical integration (where operator owns infrastructure),
- vertical separation (where operator does not own infrastructure).

In the case of vertical separation, it is necessary to address an issue of infrastructure management. The infrastructure can be managed by [3]:

http://dx.doi.org/10.5772/64820

- operator or
- authority organises infrastructure management separately from operator who provides passenger transport services.

In case that the transport operator manages infrastructure, a combination of ownership and usage may result in: [1, 4]:

- Delegated management where the operator acts independently from the authority and uses the assets provided by the authority. These assets may be provided based on various arrangements, for example "for free" or based on a contract regarding publicly owned infrastructure or through a leasing company.
- Public management where the assets are owned by the authority and transport services are provided by a public operator. Such provision of transport services may be ensured based on an in-house contract.
- In the third case which represents operators who provide assets. they are also responsible for service operation by using these assets. The contract based on which services are provided may take various forms and scope ranged from simple bus service contracts (where operators provide bus services by using their own buses) to more complex contracts (e.g. DBOT contracts—Design, Build, Operate and Transfer). Other type of contracts may represent infrastructure-concession contracts where operators may decide to some extent about the design of the assets and service realisation or PPPs contracts (Public–Private Partnership).

1.2. Public service procurement process

The process of public service procurement is complex and consists of several procedural steps, which must be done from authorities' position (see **Figure 1**) [3]. The basis of each process should be sufficient preparation. Good preparation can bring quality in services provided and effective use of public funds. Therefore, this part of the procurement process cannot be underestimated from the position of a public authority. In the first step, it is necessary to set the basic strategic objectives based on identified requirements of the public. Further, the services to be provided are characterised and designed. The services should be defined with respect to the criteria by which it will be possible to evaluate to what extent (range) a candidate (public service operator) is able to fulfil the provision of transport services. Last but not least, it is important to develop framework conditions of public service contract. Further step is related to award procedure, which can represent the direct award of contract or a competitive tendering. In the last step, a public authority implements the control procedures during contractual period whether a service operator meets its obligations.

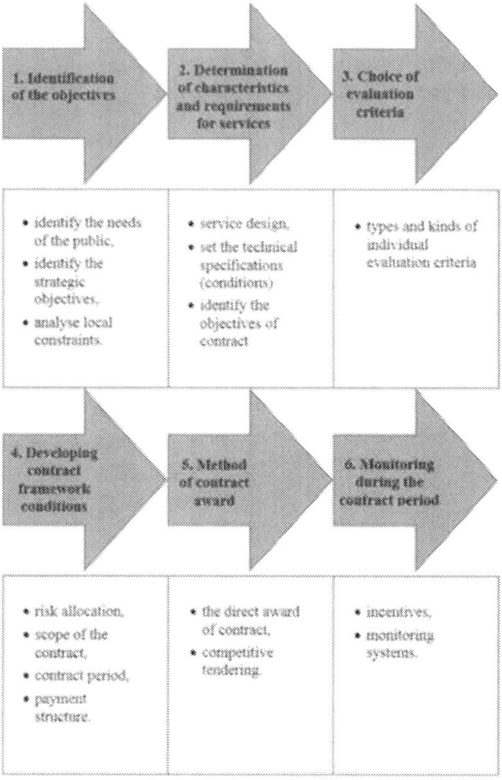

Figure 1. Procedure of public authority while contracting [3].

2. Identification of the objective

2.1. Identifying the needs of the public

The first essential step in the procurement is to determine the demand for public services—public passenger transport. Public service requirements can be found out by using different methods. One of the methods is an interview with passengers or potential passengers. The person responsible for obtaining information firstly selects a group of citizens who will be asked to answer simple questions. Ranking of requirements for public services in terms of quality and quantity is compiled at the end based on the answers [5].

Another method of identifying requirements of passengers is a form of questionnaire. The first step in creating a questionnaire is to determine the existing problem in public services.

http://dx.doi.org/10.5772/64820

Information about the problem can be obtained based on written complaints or verbal manifestations of citizens. Further, it is necessary to determine the objectives and indicators of questionnaire and to select a group of people to be queried. The advantage of questionnaires is fast obtaining the information in a short time and at relatively low costs. The questionnaires are sent directly to citizens or may be in the form of anonymous questionnaires.

By gathering information on requirements for public services, the strategic objectives can be determined.

2.2. Identifying the strategic objectives

The transport policy addresses public passenger transport and its objective is to ensure the sustainable development of mobility, coordination of public passenger transport with individual transport and improvement of road safety. Regarding the provision of public passenger transport, it is also necessary to take into account other policies such as social, environmental policy.

Typical strategic objectives within public transport are [3, 6]:

- Transport policy:
  - enhance total transport situation,
  - reliability of services,
  - ensure mobility,
  - increase market share of public transport within the intermodal market: influence the modal split, for example also by parking policy,
  - traffic safety,
  - link individual with public transport.
- Social policy—support for specific target groups:
  - people with low incomes,
  - people with limited mobility,
  - employees of the operator,
  - pupils, students and apprentices,
  - young and elderly,
  - accessibility for all layers/generations of the population.
- Environmental policy:
  - efficient energy use,
  - quality of life in urban areas,

  - reduce emission of pollutants, for example reduction in global warming gas emissions,
  - noise reduction,
  - protection of vulnerable rural areas.
- Structural and economic policy (regional development)—enhancing services within specific areas:
  - infrastructure policy—establishing capacities, regulations for use and financing the public transport infrastructure,
  - regional structure,
  - support for small- and medium-sized enterprises,
  - site-related factors,
  - location trends,
  - land-use policy.
- Budgetary aspects.

Subsequently, the strategic objectives should be confronted with the local conditions.

2.3. Analysis of local constraints

To be able to properly transform strategic objectives into tactical means (service concepts), the relevant local circumstances (constraints) have to be identified and taken into account. Numerous aspects, tasks and competencies can have an impact on public transport.

Firstly, it is appropriate to analyse the local organisation of a concerned territory and to gain a good overview of the current distribution of tasks, competencies and responsibilities between the operators and the public authorities. Some further aspects to analyse are legal and economic aspects and the existing market structure as well as the existing transport system and geographical aspects [3].

Significant local circumstances (constraints) include [4, 6]:

- Existing local organisation of public transport:
  - localisation of information and skills,
  - localisation of decision-making powers for policy-making (strategic level) service design (tactical level) and operational decisions (operational level),
  - identification of roles and duties of public authorities and service operators.
- Legal restrictions:
  - existing contractual regulations,
  - existing awarding and contracting procedures,

- right of initiative,
- overcompensation,
- national/local legal framework,
- EU legal framework.

- Economic restrictions:
  - financial/budgetary aspects,
  - ability to bear risk by the authority,
  - economic situation of the operator market (including ability to bear risk).
- Market structure of service operators:
  - capabilities,
  - number and size,
  - efficiency,
  - ownership.
- Existing transport system:
  - vehicles,
  - network design,
  - infrastructure, for example existing railways, existing depots,
  - existing databases, for example passenger data, modal split figures,
  - level of quality of public transport services.
- Spatial/geographical restrictions.

3. Determination of characteristics and requirements for services

The further step prior to the conclusion of public service contract is a characterisation of services and definition of requirements for services. It is also necessary to divide the tasks, competencies and responsibilities between the operators and the public authorities in order to ensure efficient provision of public service.

3.1. Service design

In order to reach the decision on an adequate level of competence of service operator when designing services, it is recommended to answer the following preliminary questions first [5]:

- Which interests are in conflict with each other?
- Which interests need to be harmonised (and how)?
- To what extent can a natural overlap between the commercial and other interests of the operator, the interests of the public as well as the interests of the authorities be expected?

It is necessary to note that the allocation of responsibilities determines the appropriate risk allocation between the operator and the authority. Whoever takes the opportunities and risk is the party most appropriate to influence the corresponding features.

Decision-making on service design (i.e. tactical decisions) can be organised in different ways. It should be distinguished between two basic periods:

- The period during which the contractual relation between operator and authority is established.
- The period during which the contract is realised.

For each of these two periods, fundamental organisational decisions have to be taken as to the allocation of initiative power and decision power to the authority and to the operator [3].

Within the first period of establishment of the relation, service design can be determined [2]:

- by the operator *through the bid* that he delivers to the authority; in the context of awarding, this is also known as *"functional"* awarding,
- in a negotiated way between the operator and the authority *during* the contracting process; this intermediate way to organise things is also known as *"negotiated"* or
- by the authority *prior* to contracting; in the context of awarding, thesis also known as *"constructive"* awarding.

During the second period of contract realisation, service redesign can also be organised in different ways [1]:

- It can be determined by the authority, or
- It can be determined by the operator:
  - the operator may only have the possibility to suggest amendments to the network, whereas the authority remains in charge of deciding upon the implementation of those changes after conducting a check on the desirability and/or financial consequences of the change or
  - the operator may have the freedom to modify services autonomously as he wishes (indeed, within specific norms of network accessibility specified by the authority within the contract).

3.2. Determination of technical specifications (conditions)

Technical specifications and conditions are understood to be the determination of characteristics and requirements, which must be fulfilled in the bids submitted by tenderers in order to

http://dx.doi.org/10.5772/64820

obtain a contract. Technical conditions are part of the tender documentation. Technical conditions may be determined either by the form of references to the documents, standards, regulations and acts or by the form of requirements for parameters of expected utility, for example setting requirements for performance, capacity.

Individual characteristics and requirements for services must be set so that none of the tenderers and candidates is discriminated and the principles of transparency, economy and efficiency must be applied.

3.3. Identifying the objectives of contract

The objectives should not be specified too general. It is favourable if the concretisation of subject of the contract in the form of objectives is a part of tender documentation. This contributes to the improvement of evaluation process, award procedure as well as control process. Correct determination of objectives is an assumption for easier definition of evaluation criteria. Requirements for the characteristics of objectives are presented in **Table 1**.

Characteristic of objectives	Description	Example
Verifiability	An objective can be verified	Technical parameters
Quantifiability	An objective can be measured	Number of kilometres travelled, hours of operation
Objectivity	An objective relates directly to the purpose of contract	Quality of the carriage of passengers
Consistency	Mutual continuity of objectives	Reduction the transport impact on the environment by promoting public transport

Table 1. Requirements for the characteristics of objectives.

4. Choice of evaluation criteria

A public authority must determine evaluation criteria, which reflect the expectations that are to be achieved through public transport services. The evaluation criteria can be distinguished in terms of type and kind. Individual types and kinds of evaluation criteria with stated examples are presented in **Table 2**.

4.1. Transformation of qualitative criteria into quantitative criteria

In order to determine to what extent the feature of quality is fulfilled, it is necessary to find a way to measure particular feature of quality. Transformation of qualitative criteria into quantitative criteria is performed due to the measurability of quantitative criteria. The intensity of quality can be measured, for example, by using ten-point scale. Interval of scoring is compiled from the unsatisfactory quality up to the perfectly satisfactory quality [7].

Types of evaluation criteria	Kinds of evaluation criteria	Examples
Quantitative	*Cost*	The lowest offer price
		repairs and maintenance
		operating costs
		return on investment
	Utility	Technical level
		technical parameters
		environmental impact
	Time	Interchanges
		continuity
		time of transport
qualitative	*Quality*	Safety and comfort
		caring for passenger

Table 2. Types and kinds of evaluation criteria.

5. Preparation of contract

Basically, the forms of individual contracts differ in an allocation of risks between contractual parties and the resulting structure of payments.

5.1. Risk analysis in public passenger transport

There is a methodology, not only in the SR, but also in other countries, on the basis of which the reasonable profit is determined as percentage of economically justified costs. But in this case, the operators are not motivated to save up the costs and it is also contrary to the policy of the European Union. The reasonable profit for services, which are provided in public interest, should be based on the risk assumed by operator. Therefore, an analysis of the risks existing in providing public transport services is needed.

The risks existing in providing public transport services can be categorised [2]:

- *Systematic risks*—such risks include political risks (government decisions, changes in government policy…), international risks (changes in foreign exchange rates…), economic risks (price development, population purchasing power…), interest rate changes, inflation risk and risk of unforeseen events.

- *Unsystematic risks* are the risks associated with the revenue of company and its ability to cover liabilities. These risks may be influenced by an investment project quality, deployment environment, qualified management, etc.

Existing risk can be also further divided into two groups—cost and revenue risks (see **Figure 2**).

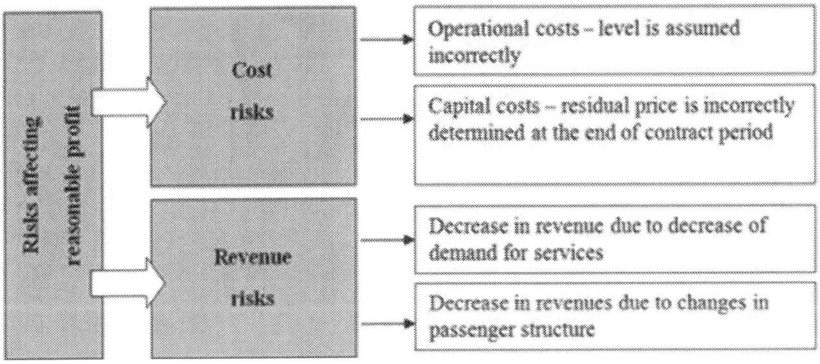

Figure 2. Classification of the risks in public passenger transport.

5.1.1. Cost risks

The cost risks are associated with a cost calculation when contracting in public economic interest. In public service contract, it is necessary to agree on a price for realised performance, which consists of the costs and profits of public service operator. In the case that the operator assumes the cost risk, it is necessary to agree on a scope of realised performance for the contract period and economically justified costs per unit of the realised performance between operator and authority. The cost risks can be divided into two groups [5]:

- *Operational cost risks* which are related to the difference of the anticipated costs calculated and the actually observed costs after performance realisation. The reasonable profit must depend on an allocation of this risk. When the operator does not assume the risk and after realisation of performance he proves eligibility of costs to authority for the purpose of compensation, the operator takes no cost risk for the performance realisation. In the case that the agreed unit costs in public service contract are final, the operator assumes the cost risk and this should be reflected in appropriate level of reasonable profit. The operational cost risks can be further divided as follows:
  - *External operational cost risks*—the risk that cannot be influenced by the operator at all (e.g. cost increasing due to flooding streets in the event of natural disasters). This group can also include the risk which can be influenced by operator indirectly or only in small extent (e.g. changes in energy prices during the contract period, change in employees' costs, etc.).
  - *Internal operational cost risks*—the risk that can be influenced by the operator, for example the costs of maintaining of vehicle fleet (the operator can decide on the maintenance process in order to avoid failure of vehicle and higher costs).
- Investment cost risks are related to the difference of the anticipated life of the fixed assets of the operator. While providing public passenger transport, it is primarily the means of transport and infrastructure (e.g. bus and tram stops, tram tracks). The reasonable profit

must depend on which party assumes the risk of the difference of actual net book value of fixed assets at the end of a contract period compared with anticipated net book value.

5.1.2. Revenue risks

The revenue risks are associated with the difference between expected revenues from operation of public passenger transport and actually achieved revenues at the end of contract period. These risks may be taken either by authority or operator and in this regard there must be appropriately set a profit level of the operator. When the authority assumes the revenue risk, a contractual relationship between the authority and the operator, which sets a compensation for realised performance, is based on the following formula [6]:

$$C = (UC + RP).P - R \tag{1}$$

where C—compensation of the authority for the operator,

UC—costs per unit of realised performance,

RP—reasonable profit for the operator expressed per performance unit,

P—the realised performance,

R—revenue achieved when realising performance.

When there are agreed final costs per unit in public service contract, which cannot be changed during a contract period, the cost risks are fully borne by the operator. The revenue risks are borne by the authority. This means that if operator's revenue is decreasing, the compensation from authority's party is increasing.

When the operator assumes the revenue risk, in the contract there is determined in addition to realised performance also absolute amount of compensation, which cannot be changed during a contract period. The compensation is based on anticipated costs and revenue, while changes in costs and revenue pose a risk of the operator. A part of the compensation is a reasonable profit of the operator resulting from cost and revenue risk of realised performance.

The cost risks are not usually related with interventions of public authorities (with an exception of changes in tax burden of the operator), and currently, they are usually transmitted to operators. In the case of revenue risks, it is possible to define influence of public authorities on revenue risks; the risks can be divided into two groups:

- *revenue risk associated with a decrease in demand*—it is a risk related to the changes in number of passengers carried when providing public passenger transport. In the case that the authority bears the revenue risk, it is necessary to appropriately involve the operator in compliance with required quality because the amount of the compensation in this case does not depend on the number of passengers carried (In the SR, this risk is very significant because the demand for public passenger transport expressed in passenger kilometres (pskm) is decreasing annually in road and railway transport). When it comes to the revenue

http://dx.doi.org/10.5772/64820

risk associated with a decrease in demand, it is necessary to distinguish territories in which the transport services are operated. The development of number of passengers carried depends to some extent on the interventions of public authorities, which can indirectly influence the number of passengers carried through a fulfilling their strategic objectives which can be economic, environmental, social and governmental.

- *revenue risk associated with a change in passenger structure*—it is the risk of revenue change because of a change in passenger structure. For example, when the selected groups of passengers (students, pensioners) travel with special fares, an increase in number of those passengers while keeping the total number of passengers causes a decrease in total revenue for providing transport services. The good solution is setting an appropriate pricing policy of transport services. However, it is important to monitor the impact of price changes on the demand, which varies considerably for particular groups of passengers. In the Slovak Republic, the discounted fares known as saver tickets (half price of a full fare ticket) are for young people aged 6 to 15 and students to 26, and fares known as "other fares"[1] are for: senior citizens over 70 (€ 0.20 per every 50 km, severely disabled people (half fare travel), parents travelling to visit their physically or mentally disabled, chronically ill children nourished in special facilities in Slovakia (half fare travel). The public passenger transport fare is regulated by public authorities that decide which specific groups of passengers will be entitled to reduced fares; and therefore, the revenue risk associated with the change in passenger structure can be classified as the risks associated with interventions by public authorities.

5.2. Risk allocation between the contracting parties

The authority has to decide upon how to allocate risks between contracting parties appropriately. Risk can have a negative effect on the outcome of contracting, especially when using competitive awarding [3]:

- The higher the risk, the lower the number of bidders (high entry barriers).
- A very high level of risk, resulting out of a high level of uncertainty, may result in a higher danger of insolvency for operators in case of a full realisation of the risk.
- The higher the risk, the higher the risk premium the operator is calculating (increasing the subsidy to be paid by the authority).

Therefore, from a very schematic point of view, risk can be classified as follows [1]:

- *High risk*—High uncertainty and/or critical for operator in case of realisation (operators will calculate a high risk premium).
- *Low risk*—Predictable for operators and/or not critical for operators in case of realisation (operators will calculate a low risk premium).

---

[1] Decree of Office of Rail Regulation No 654/2005 lays down the scope of price regulation for railway transport and price quotations of self-governing regions which determine the maximum prices for national regular bus transport when the distance from origin to final bus stop exceeds 100 km

- *Unbearable risk*—Unpredictable and critical for operator in case of realisation (risk not bearable for operators—market entry barrier).

A contract form and payment structure is dependent on a way of the risk allocation. The contract forms are as follows:

- *Management contract*—operator bears no risk; cost and revenue risk is borne by authority that pays the economically justified costs to operator. Those costs are accounted in the end of period. This means that the risk from difference between anticipated and actual costs is borne by authority, which bears also the risk from difference between anticipated and actual revenue. In this case, the level of reasonable profit of operator should relate only to numb capital during providing transport services because he bears no risk. The reasonable profit, in this case, must include, in addition to the numb capital, also a reward for assuming the cost risk.

- *Gross cost contract*—operator bears cost risk; the risk from difference between anticipated and actual costs in the end of period and the authority bears the risk from difference between anticipated and actual revenue.

- *Net cost contract*—operator bears cost and revenue risk. In this case, the operator bears the risk from difference between anticipated and actual costs/revenue, which are identified in the end of contract period. The authority pays only compensation, which is agreed before realised performance to operator. This means that the authority bears no risk.

The essential advantages and disadvantages of gross and net cost contracts are presented in **Table 3**.

	Advantages	**Disadvantages**
Gross cost contract	• Reliable calculation for operator. • Low barriers for market entry. • High legal certainty. • Enforcement of public interests (e.g. tariffs). • competition neutrality.	• No entrepreneurial interest. • High level of regulation. • No budget reliability for public authorities. • High monitoring efforts.
Net cost contract	• Preservation of entrepreneurial interest. • Lower monitoring charges—customer as "adjustment factor". • Constant amount of compensation payments. • Minimal requirements have to be defined.	• Low legal certainty (in case of tendering because of missing database). • Higher risk for operators—risk premium and higher compensation payments. • No incentives to consider social, environmental and political goals.

Table 3. Advantages and disadvantages of gross and net cost contracts.

5.3. Determination of risk and reasonable profit when financing public passenger transport

Determination of reasonable profit as a percentage of costs is economically incorrect in a regulated sector. The reasonable profit must relate to the risk that is borne by operator in regard to realised performance. This means that the level of reasonable profit must be higher in case of the operator bearing cost and revenue risks in comparison with the operator bearing only cost risks while keeping the same range of performance.

Based on the previous analysis of risk allocation, the level of reasonable profit can be defined as follows [1]:

1. *Operator bears no cost or revenue risks*—the risks associated with providing transport services are borne only by authority; and therefore, the level of reasonable profit should relate only to the capital used by operator when providing transport services. A reward for provided capital of operator should depend on profitability level of capital invested in terms of deposits with guaranteed returns. The reasonable profit in management contracts is calculated according to following formula:

$$\mathrm{RP} = \mathrm{CO.p} \tag{2}$$

where: RP—reasonable profit,

CO—capital invested by operator for providing transport services,

p—capital profitability.

2. *Operator bears cost risks*—the level of reasonable profit must consist of two parts: the reward for provided capital of operator (the same as mentioned above) and the reward corresponding to the cost risks. The reasonable profit when contracting for public interest and where operator bears cost risk is possible to determine according to following formula [6]:

$$RP = CO.p + \left(\sum_{i=1}^{n}(C_i.R_{Ci})\right).P \tag{3}$$

where: C_i—i's value of cost item of operator in unit expression,

R_{Ci}—risk of assumed value of i's cost item in percentage expression from cost item value,

n—number of operator's cost items,

i—i's cost item of operator,

P—realised performance.

It is necessary to define the way of risk determination of estimated values of individual cost items in relation to reasonable profit. The risk can be calculated by using the relationship for determination of safety surcharge to net premiums.

3. *Operator bears cost and revenue risks*—the level of reasonable profit must consist of three parts: the reward for provided capital of operator (mentioned above), the reward corresponding to the cost risks (mentioned above) and the reward corresponding to revenue risks. The reasonable profit when contracting for public interest and where operator bears cost and revenue risks is possible to determine according to following formula [6]:

$$RP = CO.p + \left(\sum_{i=1}^{n} (C_i.R_{Ci}) \right).P + \left(\sum_{j=1}^{m} (R_j.R_{Rj}) \right).P \tag{4}$$

where j—j's group of passengers with the same fare level,

m—number of passenger groups which are different by fare level,

R_j—assumed revenue of j's passenger group in unit expression,

R_{Ti}—revenue risk of j's passenger group expressed in percentages.

Determining revenue risk is done by an analogous method such as in case of determining cost risk. Revenue risk is possible to determine at standard deviation level of income change per individual groups of passengers in observed period.

5.4. Scope of contract

In terms of the scope of the contract, a public authority can decide on:

- *route contracts*—used for a specific bus line or can include a group of shorter bus lines located close to each other,
- *network contracts*—these contracts cover whole city territory and network of city public transport or they are related to more transport modes such as metro, bus and tram,
- *sub-network contracts*—related only to a certain part of city (e.g. suburb of city) and only one mode of transport.

The contract size has an influence on efficiency. The contract size has an influence on efficiency. Related to matters of risk, this aspect mainly affects the market entry possibilities and might result in an overly elevated complexity level for the respective service operators.

Network contracts [3]:

- increase the need to select long term contracts,
- provide integrated public transport services delivered by one operator to passengers,
- produce market entry barriers for small- and medium-sized companies,
- account for a great operational complexity,
- enable net cost contracts,

- provide substantial optimising opportunities to the operator and therefore may increase efficiency levels,
- might be more difficult to monitor.

Route contracts [3]:

- low market or no market entry barriers exist for small- and medium-sized operators,
- integration of public transport services needs to be realised through other organisations (authority or related body),
- in case of dependency on the performance of other operators, net cost contracts are not recommended,
- provide fewer optimising opportunities.

Sub-network contracts

Sub-network contracts provide a compromise between network contracts and route contracts if required.

5.5. Duration of contractual period

When designing the length of contract period, the public authority should take into account the level of revenue risk borne by the operator in order to allow him to develop market activities for increasing the number of passengers.

5.5.1. Flexibility during the contract period

Changes in external factors, political aims or passenger needs can lead to a need for amendments to service design during contract period. Therefore, certain flexibility should be incorporated in contracts.

Contracts should contain appropriate variations and termination clauses [3]:

- Enable service redesign by the operator autonomously when using net cost contracts (with functionally designed minimal standards) while preventing negative financial impact to authority.
- Enable the authority and the operator to terminate (or at least renegotiate) the contract in case of major unforeseen changes with major commercial influence.
- Check whether there will be major changes during the contract time (e.g. a new bus lane within the centre during the contract period) and insert suitable agreement procedures on how to deal with these circumstances.
- Enable service redesign by the operator after approval of the authority under all awarding models, based upon fixed price list and limitations (e.g. limited increase in vehicle-km) to reduce risk for the operator and the authority.

- Insert arbitration clauses to avoid unproductive conflicts.
- Enable service redesign by the authority in case of constructive design, based upon fixed price list and limitations (e.g. limited increase in vehicle-km) to reduce the risk for the operator (and the authority).

It is necessary to note that the longer the contract period, the more increases the need for flexibility of the contract. In case of high uncertainty about future developments (e.g. major changes within the coming years without any sufficient expectations on the influence on the contractual outcome), a short contract period is recommended (maybe including extension options) [1].

In terms of decision-making on contract period, it can be recommended as follows [5]:

- Decision based on trade-off between flexibility (short-term contract) and increasing incentive to make capital investments (long-term contract).
- Use of short-term contracts in case of high uncertainty about future development (e.g. net cost contract with high uncertainty about development of the ridership).
- Avoiding too short contract periods as this causes increasing uncertainty (which may result in lower interest of operators on that contract).
- Use of long-term contracts in case of high specific investment needs with long amortisation periods, including review dates on the performance.
- Avoiding too long contract periods in case of competitive awarding to secure competition within the market.
- Use of longer-term contract when substantial market action is required from the operator (take account of longer lead times to develop measures and to reap the profit of their implementation).
- Use of short-term contracts in case of the need for increased flexibility.
- Avoiding too long contract periods to be able to recalibrate contract clauses according to market development.

5.6. Payment structure

Payments which are paid to operators in return for service provision may represent variable, fixed payments or their combination. However, it is important to determine a clear and verifiable payment structure in order to avoid misunderstandings and disputes between contracting parties during the contract period. It should be also noted that financing of infrastructure should be separated from operation financing due to transparency reasons [3].

Payments may flow not only from an authority to an operator but also in opposite direction from an operator to an authority. The situation is dependent on the market conditions. For example, the award of a very profitable service contract to the service operator through a competitive tendering may bring the situation of payments flowing from the operator to the authority. However, as many public transport services cannot be provided on a commercial basis, the payments which are paid to service operators from authorities are more usual in practice. The situation also depends on the structure of the additional incentives, which may be included in individual public service contracts.

The amount of the payment to be paid depends on various factors. These factors relate to the type and scope of a particular public service contract, which is awarded to a particular operator [1].

To reduce the risk level, it is recommended to include a lump sum payment. Inclusion of variable payments, which represent incentive-based payments into a public service contract, may motivate the operator to achieve the objective set by the authority.

As previously mentioned, the structure of payments must be determined in a clear manner. It is recommended to reduce complexity as much as possible in favour of a simple payment structure. This may also reduce the level of risks and avoid the high entry barriers to the market.

6. Procedure from the position of service providers in procurement of the transport service

In general, the candidates (tenderers) are understood to be the entities who offer a solution of bids (public service contract). Transport services may be provided after the conclusion of public service contract between the successful tenderer and competent authority (public authority). The main objective of a tenderer is to gain a competitive advantage over the other tenderer and thus succeed in a competitive tendering.

6.1. Procedure of service providers in competitive tendering

Procedure of service providers in competitive tendering should consist of the following steps:

- analysis of own position in competitive market,
- analysis of the needs and requirements of the public,
- analysis of technical conditions, the subject and criteria of the public contract,
- compilation of the bid and its submission.

6.1.1. Analysis of own position in competitive market

Success rate of a tenderer is dependent on the quality and quantity of services offered. The tenderer as an economical subject acts as a competitor in relation to other tenderers and towards the contracting authority as a potential provider of services. The tenderers, who wish to be involved in competitive tendering for public services for the first time, should perform an analysis of the competitive environment in a given sector of services. This analysis includes an analysis of strengths and weaknesses, and the opportunities and threats. Thanks to the analysis, the tenderers may find out their position in the market or look for ways to improve or retain that position.

In the case that tenderer already participated in competitive tendering in previous periods, he may have plenty of information for predicting the capabilities and behaviour of the competitors. The tenderer should already be familiar with the evaluation criteria as well as their weights of importance. He should be able to create his own evaluation system, the results of which are depicted on the matrix of strengths and weaknesses. On this basis, it is then possible to evaluate own chances of tenderer or take action to improve existing conditions.

6.1.2. Analysis of the needs and requirements of the public

In the case that tenderer is able to identify the exact needs of the public, he has an advantage over the competition. The first step is to identify deficiencies in area of public passenger service, for example number of joints, accessibility of stops, points of transfer, continuity. According to these findings, a tenderer is able to create the concepts and plans for increasing the number of passengers in the future, which may increase the likelihood of success in a competitive tendering.

6.1.3. Analysis of technical conditions, the subject and criteria of the public contract

In terms of technical conditions of contract, a potential provider of transport services must examine in detail the technical conditions, characteristics and requirements for transport services.

Technical conditions may be distinguished as follows [2]:

- technical conditions in relation to expected fulfilment of public contract,
- technical conditions promulgated by contracting authority.

Precise definition of the subject of public contract is a certain orientational point to obtain the public contract by a tenderer. Subject of fulfilling the public contract is defined in the tender documents. In the case that tender documents do not contain the subject of public contract, a contracting authority is obligated to send to a tenderer the subject of public contract in written or electronic form within the period stipulated by the Act.

6.1.4. Compilation of the bid and its submission

Content of the bid should be drafted so that a tenderer is able to demonstrate:

- financial position,
- technical competency,
- professional competency.

The tenderer is obliged to draw up a bid based on the instructions for compiling bids that are contained in tender documents. He is obligated to comply with all the requirements specified in those instructions in order to succeed in competitive tendering.

It would be appropriate from tenderer's position to nominate the person responsible for compilation and submission of bids.

6.2. Approach of operators to various contract forms

From the position of public authorities that plan funds for providing public transport services, the net cost contracts appear to be the most advantageous. Under this contract form, all the risks, cost and revenue, are borne by the operator. The authority pays to operator a financial amount that is fixed determined at the beginning of a contract period and stated in the contract. In this case, the public transport services in a given area are provided only by selected operator through a license. Such an operator has the option to set the level of fares because he also assumes revenue risks.

The gross cost contract is advantageous for operators because they do not bear the risk of revenue decreases, which is usually associated with the factors that cannot be influenced by operators.

Based on mathematical modelling of a price regulation and determination of business reasonable profit in network industries, Fendekova and Fendek [8], they mathematically model an approach of the enterprise in regulated sector and they define two approaches that can be applied in providing public transport services:

- Approach of enterprise applying return on investment—the approach encourages an enterprise to use a high volume of capital in order to achieve the maximum permitted reasonable profit. The enterprise has no incentive to use more efficient combination of inputs, for example supporting employment in comparison with an end in itself investment in facilities.
- Approach of enterprise applying increasing the volume of outputs—in this case, if the authority does not have the possibility or manpower for verifying effectiveness of providing public transport service, the operator will seek to realise also inefficient performance.

6.2.1. Approach of operators to gross cost contracts

The operator assumes all cost risks under gross cost contracts in providing public transport services, whereas the authority bears revenue risks related to a decrease in the number of passengers. Documents for optimisation of public transport services are available for the operator and in case that the authority does not have sufficient access to the data about the number of passengers on particular bus routes, he is not able to optimise public transport services. It is necessary to continuously optimise providing public transport services when the number of passengers decreases. In terms of business interest, the operator who bears no revenue risk is willing to operate also the buses without any demand because the authority bears the risk that bus will not be used by passengers. For example, if there was abolition of a production plant into which the operator provides transport services for employees and the authority did not change a transport license, the operator would continue in providing transport services because a decrease in revenue (in this case to the zero level) would be compensate by the public authority assuming revenue risk.

This approach assumes that a fare level is also determined in the public service contract. The deficiencies of such contracts may be addressed by contractual clause based on which the authorities have an access to the electronic data on the number of passengers in real time and thus they can obtain materials to optimise the transport services.

6.2.2. Approach of operators to net cost contracts

Under these contracts, the operators assume not only cost risk but also revenue risk related to providing transport services. The authority grants a license for providing public transport services to the operator that is then entitled to provide public transport services in a given served area with an exclusion of other operators (during the license period). Following from the analysis processed by van de Velde [3], the net cost contracts are rarely awarded as route contracts because the operator determines a fare level and he becomes a monopoly for providing public transport services in a given served area during the licence period. The following mathematical model defines a procedure of such operator in relation to providing transport services.

Assume that the operator is a company that aims to make a profit. Based on a license and a public service contract—net cost contract, the operator provides a range of transport services bounded by demand of q. Start from a general assumption which is acceptable in any type of market structure, the consumption of a product offered in the market is described by a price-demand function that expresses willingness of consumer to buy q units of services provided at given price—p.

$$p = p(q) \tag{5}$$

Technological conditions of the operator are expressed through the real cost function:

$$n = n(q) \tag{6}$$

The equation presents the amount of minimum costs of n which are spent by producer in the production of q units of goods, while it is assumed that a price -demand function p(q) is continuous and twice differentiable real function. It is also envisaged that the price-demand function of consumer is constructed in order to clearly motivate the consumer to buy q units of services at market price—p—because the consumer feels the maximum rate of usefulness from consumer strategy realisation in this combination of price and demand. Analogously, the cost function describes a process of providing services by operator so that quantifies the minimum of total production cost—n for an optimal combination of production factors required to produce q units of provided services.

While optimal consumer behaviour is described by price—demand function p(q), the optimal operator behaviour is described by a profit function $\pi(q)$, which is formulated as the difference between revenue and costs of company corresponding to a certain production volume of q:

where a continuous and twice differentiable real function of company revenue r(q) is defined as the product of price and supply volume, that is:

$$\pi(q) = r(q) - n(q) \tag{7}$$

$$r(q) = p.q = p(q).q \tag{8}$$

A company operating in every type of market structure (a competitive company as well as a monopoly) seeks in a decision-making process such a combination of price and supply of its product that guarantees a maximum level of the profit. This means that the operator also provides transport services in such a way that ensures the maximum profit. Analytically, this approach can be expressed as follows:

$$\pi(q) = r(q) - n(q) = p(q).q - n(q) \tag{9}$$

For optimising profit function, it is necessary that the function would reach its maximum at certain point of supply—q, that is that the first derivative of the profit function at this point is zero:

$$d\pi(q) / dq = d(r(q) - n(q)) / dq = rm(q) - nm(q) = 0 \tag{10}$$

In the Eq. (10), rm(q) is a marginal revenue function of the operator and nm(q) is a marginal cost function. Based on Eq. (10), it can be seen that a company generally achieves a maximum profit for a volume of q when the marginal revenue equal to marginal costs, that is a solution to the equation:

$$rm(q) = nm(q) \tag{11}$$

Then, it is possible to calculate such a price—pp that maximises profit of the operator at the optimal level of supply qp:

$$pp = p(qp) \tag{12}$$

In the case of the operator who operates in non-regulated sector (e.g. long-distance transport), where the competition exists, the approach described in previous relationships (equations) cannot be applied. The operator accepts the price—pK at the level of his marginal costs—and he offers the production volume—qK at that price. This means that the following relationship applies:

$$pK = nm(qK) \tag{13}$$

On the other hand, a monopoly due to its dominant position in the market can influence the price of its product so that to achieve higher profit in comparison with competing companies. The monopoly determines an optimal price—pM based on the optimisation solution Eq. (9) and based on relationships Eqs. (11), (12), that is:

$$pM = p(qM) \tag{14}$$

Based on above mentioned, the operator operating in a monopoly position can provide fewer services at a higher price compared to competitors. The approach is shown in **Figure 1** based on which the following applies:

$$pM > pK \wedge qM < qK \tag{15}$$

It can be concluded based on **Figure 3** that the operator operating in a competitive market would provide services in a volume of qK at the price—pK. If the average unit costs per unit of provided services are defined as:

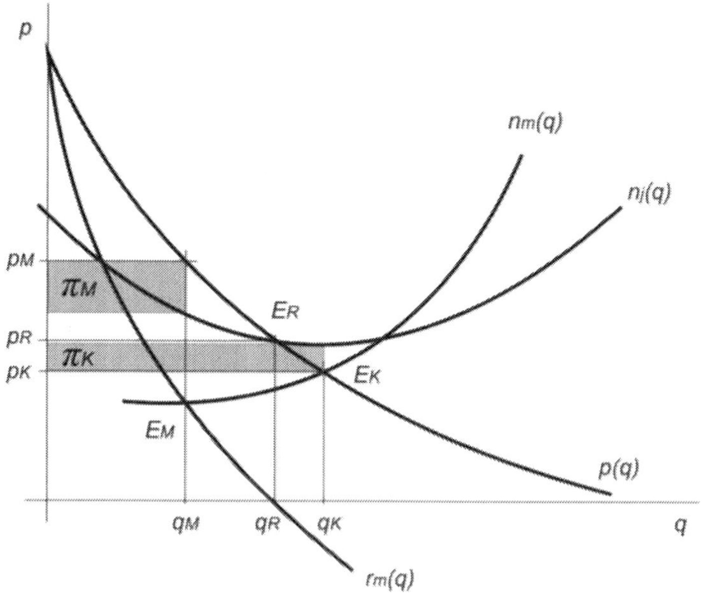

Figure 3. Monopoly and competitive company.

$$nj(q) = n(q) / q, q > 0, \tag{16}$$

Then the price of provided services will not cover even the average costs of the operator because:

$$nj(q) > pK \tag{17}$$

If the operator provides public transport services in such a case, the loss of operator will be at the level of (according to the Eq. (9)):

$$\pi K = rK - nK = pK.qK - njK.qK = (pK - njK).qK, njK > pK \tag{18}$$

If the operator acted as a monopoly in the same market, he would provide public transport services at the level of qM at price—pM and he would achieve, under these conditions, a profit —πM at the level (**Figure 3**): because the following applies for the monopoly:

$$\pi M = rM - nM = pM.qM - njM.qM = (pM - njM).qM \tag{19}$$

$$pM > njM \tag{20}$$

While providing public transport services, the operator in a monopoly position achieves higher profit in comparison with the operator who operates in the market of perfect competition. If the public authority decides on a net cost contract, according to which the providing transport services is in the competence of the operator, there will be the risk of lower quality or the risk of lower performance than in comparison with the case of gross cost contract.

The public authorities tend to issue a license for one operator to provide public transport services for whole served area and consequently to conclude a net cost contract. Under this contract, the decision on an organisation of public transport service including pricing is in the competence of the operator. However, it is important to note that this procedure can lead to reducing quality of providing services.

Author details

Miloš Poliak

Address all correspondence to: milos.poliak@fpedas.uniza.sk

University of Žilina, Žilina, Slovakia

References

[1] Poliak, M. Vzťah primeraného zisku a rizika v hromadnej osobnej doprave na Slovensku. Journal of Economics. 2013;61(2):206-220.

[2] Stanley, J., van de Velde, D. Risk and reward in public transport contracting. Research in Transport Economics. 2008;22(1):20-25.

[3] van de Velde, D. et al. Competitive tendering in the Netherlands: central planning vs. functional specifications. Transport Research Part A. 2008;42(9):1152-1162.

[4] Lalive, R., Schmutzler, A. Exploring the effects of competition for railway market. International Journal of Industrial Organization. 2008;28(2):443-458.

[5] Poliak, M., Krizanova, A. Semanova, S., Stefanikova, L. The influence of contract form choice of the transport services ensuring on performance contracting entity requirement. Transport Problems: International Scientific Journal. 2014;9(4):153-161.

[6] Hensher, D.A. - Wallis, I.P. Competitive tendering as a contracting mechanism for subsidizing transport. Journal of Transport Economics and Policy. 2005;39(3):295-321.

[7] Nedeliakova, E. - Panak, M. New trends in process-oriented quality management. Procedia Economics and Finance. 2015;34(1):172-179.

[8] Fendeková, E., Fendek, M. Modely cenovej regulácie sieťových odvetví. Journal of Economics. 2010;58(10):1039-1054.

Chapter 9

Financial Aspects of Urban Transport

Mihaela Grubišić Šeba

Additional information is available at the end of the chapter

http://dx.doi.org/10.5772/65751

Abstract

The demand for urban transport quality is typically above financing possibilities of public authorities. Financing urban transport has always been one of the prime problems of city authorities because of the necessity to connect the city centres with their surroundings and to enable time saving and thus better quality of life for the citizens. This problem especially emerges in the twenty-first century when the citizen requirements for better urban and suburban connectivity are coupled with smart, intermodal and energy-efficient urban transport. The financing problem of urban transport is somewhat simpler in very populated and developed areas as the growing number of public transport users continuously finance urban transport fleet renewal. However, less developed areas have to have integrated pricing and social policies towards the end users of urban transport, which often turns to be unsustainable in the longer period of time. Depending on the project size, financial strength of municipalities and/or central state, urban transport infrastructure construction and maintenance are typically financed from national or local state funds or borrowing. Some urban transport lines can also be given in a concession. Financing urban transport encompasses either financing urban transport infrastructure construction or financing fleet renewal, or combined financing of both urban transport infrastructure and fleet renewal. The EU funds have contributed much to financing urban transport needs, especially in large metropolitan areas. Yet, many countries opt for financing regional and cross regional connectivity by roads, rail, airports or waterways, while urban transport remains a care of national or local public authorities. Most literature is devoted to rail, road and port infrastructure construction in general, while urban transport fleet renewal and operating performance of urban transport operators have not been widely discussed. This chapter aims to partly fill in this gap for the selected cities of formerly planned economies of the Central and Eastern Europe and Southern and Eastern Europe.

Keywords: urban transport, operating performance, fleet financing, transition economies

1. Introduction

The task of all public authorities is to insure the provision of safe, reliable and smooth-running (collective scheduled) passenger public transport in the urban and suburban city area. About 185 million passengers on average across the EU capital cities used the city public transport [1] in 2014, which is on average three rides per citizen per week or 150 rides per year. The statistics lacks for other cities, however. Of 57.9 billion journeys within the EU capital cities in 2014, 55.8% public transport passengers were transferred by buses and trolleybuses, 14.5% by tram, 13.6% by suburban rail and 16.1% by metro [1]. People in some cities reach desired locations in very reasonable time and do not need to have a car for urban transport purposes, while people from other urban areas travel by public transport so long that they appear to be much farer from some city areas than it is actually the case. Most people still travel between 30 and 40 minutes in one direction [2].

Good urban public transport organisation is crucial not only for sound city functioning but also for the urban economy as cities contribute more than 80% to overall EU gross domestic product (GDP) [1]. Public transport in the EU is governed by the Regulation 1370/2007 which came into force in the beginning of 2009 [19]. It regulates both national and international operation of public passenger transport services by rail, other track-based modes and road and public subsidies thereof. The maximum duration of public service contract is limited to 10 years for road transport and 15 years for rail transport, although it could be lengthened by 50% of the time if the operator needs to provide significant assets during the contract term. However, city transport service provision may be excluded on the grounds of their special historical or touristic interest.

Urban fleet is also subject to regulation of fuel consumption effects as transport in general, and urban transport in particular is one of the greatest causes of greenhouse gases (GHG) emissions and global warming. Most urban fleet vehicles are run on diesel fuel. Transport in Europe is almost 90% dependent on oil and its derivatives, and urban transport produces about a quarter of total CO_2 emissions from transport [3]. Other types of fuel consist of compressed natural gas (CNG), liquefied petroleum gas (LPG), biogas and biodiesel, while there are also full electric vehicles. Within the EU the fuel standards I–VI apply for all vehicles. Standard VI is the most stringent, while the differences between particular fuel standards are described in Ref. [4]. Many European cities procure only environmentally friendly buses, requiring fuel of either standard V or VI. Throughout Europe the mostly purchased buses are those of 12.2 m length, with 40–50 sitting and 30–70 standing places, with about 18 t weight. The cost-benefit trade-off (e.g. between diesel and CNG-run vehicles) is lower/higher purchase price vs higher/lower GHG emissions during the expected economic lifetime for the vehicle [5]. Due to constant technology advancements, the monetary difference between available technologies decreases. A research [6] shows that capital costs of ten diesel bus fleet purchase were 0.17 compared to 0.12 euro/km for 10 CNG bus fleet price, calculated on average 445.449 and 423.250 km passed in 2013 for diesel and CNG buses, respectively, and estimated annual depreciation. Maintenance costs of 0.28 euro/km passed favoured diesel buses as opposed to CNG buses whose maintenance cost reached 0.35 euro/km passed. However, operation costs were 0.64 euro/km

http://dx.doi.org/10.5772/65751

for diesel buses vs 0.41 euro/km for CNG buses in 2013. Overall, the analysis [6] showed the total costs of diesel bus fleet of 1.03 euro/km compared to 0.92 euro/km for CNG buses in 2013. Such a difference is based on the average fuel price of 1.11 euro/l for diesel and 0.58 euro for 1 m^3 of CNG. The EU plans to halve the use of conventionally fuelled cars in urban transport by 2030 discontinue their use in cities by 2050 and achieve CO_2-free city logistics in major urban centres by 2050, all to reduce GHG emissions by at least 80% in 2050 [3]. This ambitious goal implies greater than ever expected use of public transport in the following decades. The level of ambition is striking as currently personal cars account for 81.6% total passenger kilometres in EU-28 [7] and public transport for the rest, i.e. buses and coaches (9.3%), railways (7.4%) and tram and metro (1.7%). The separate statistics on trolleybuses is not led, although trolleybuses are strongly recommended as clean and silent mode of public transport throughout the EU. The cost of a trolleybus is about 20% in addition to the price of a standard city bus, but its expected lifetime is 20 instead of 12 years, it is more environmentally friendly, and it consumes less electric energy than buses do diesel fuel in monetary units [8]. In terms of expected economic lifetime, the highest benefit-cost ratio has metro and light rail. One metro vehicle costs on average about 1.5 million euro with expected lifetime of 40 years, whereas one tram/light railway vehicle cost ranges from 1.2 to 1.5 million euro. However, the construction of metro network is estimated to 130 million euro/km, whereas the construction of light rail/ tram network is much cheaper at about 15 million euro/km on average [9].

Regular public fleet purchase depends very much on the organisation of city public transport and operating performance of the urban transport operators. This chapter reveals the mutually linked factors that cause such differences and hence the differences in public transport fleet and overall public transport quality.

2. The organisation of the city public transport

Transport statistics is in general more often available for main transport infrastructure at country levels (roads, motorways, number of buses, number of passenger kilometres, etc.). Available public transport statistics usually combines national, regional and urban transport. The data on operation of either public or private urban transport operators are widely not available, or they are available only for metropolitan areas and some largest companies if they are not a part of the city holdings. While there are some studies explaining the organisation of urban public transport in some bigger Central European cities such as Prague, Budapest or Kraków, the data on city public transport in Western Balkan countries is widely neglected. The reason for putting them together is that they all belonged to centrally planned economies until the 1990s and followed different economic path thereafter.

The city transport operators presented in **Table 1** have been selected based on the criterion of more than 100 thousand inhabitants of the city. However, due to the large subsample size, i.e. the disproportionate number of cities of such size per country, this threshold has been increased to 145 thousand citizens for Czechia and Hungary only. **Table 1** also contains the data on transport modes available in the selected cities as well as the data on majority ownership of the public transport operators. In all cities the public authority owns the public infrastructure, while the public operator operates either its own fleet or the public fleet. The city public transport operator can be in public, private or mixed ownership, while in most cases, it is publicly owned. Some city transport operators are independent companies, while some of them are parts of holding structures (Zagreb). Public transport operators in the cities are typically the utilities owned by local authorities that have been providing such services for decades. However, some cities have contracted other operators in addition (Prague, Budapest). Such contracts are an exclusive right to use a certain route or lines or operate the entire network of urban and/or suburban lines in certain city area. Subject to predefined quality of service, such contracts may be authorisation rights (licences) with or without investment in any kind of transport infrastructure that are valid for a certain number of years. In some countries this term is bound to the depreciation period of the public transport fleet. There are also public transport operators covering entire regions and/or multiple urban and suburban areas in a certain region. The example of such a company is Arriva Dolenjska in Primorska in Slovenia, a Deutsche Bahn-owned company, providing public transport services in Slovenian cities: Novo Mesto, Koper and Piran.

Country	City	Modes of transport	Urban transport operator	(Majority) ownership of the operator	No. of city citizens as per latest census
Czechia	Prague	B, M, T, R	Dopravní podnik HL.M. Prahy, a.s.22% of public bus transport is operated by other operators Czech railways (one urban railway line)	Public Private Public	1.289.211
	České Budějovice	B, TB	Dopravní podnik města České Budějovice, a.s.	Public	154.588
	Brno	B, T, TB	Dopravní podnik města Brna, a.s. operates both urban and regional transport in South Moravia	Public	385.913
	Plzeň	B, T, TB	Plzeňské městské dopravní podniky, a.s.	Public	188.045
	Ostrava	B, T, TB	Dopravní podnik Ostrava, a.s. Arriva Morava, a.s.	Public Public (but international)	326.018
	Hradec Králové	B, TB	Dopravní podnik města Hradce Králové, a.s.	Public	145.373
	Olomouc	B, T	Dopravní podnik města Olomouce, a.s.	Public	161.641
Slovakia	Bratislava	B, T, TB	Dopravný podnik Bratislava, a.s.	Public	432.000
	Košice	B, T, TB	Dopravní podnik města Košice	Public	240.433
Hungary	Budapest	B, M, T, TB, R	Budapest public transport company (Budapesti Közlekedési	Public (the only operator until 2010) Public	1.741.041

http://dx.doi.org/10.5772/65751

Country	City	Modes of transport	Urban transport operator	(Majority) ownership of the operator	No. of city citizens as per latest census
			Zártkörűen Működő Részvénytársaság—BKV) MÁV Hungarian State Railways Private Company VOLÁNBUSZ Transport Company VT-Arriva	Public (but national) Public (but international)	
	Miskolc	B, T	MVK Rt.	Now private (a part of Miskolc holding, but it is used to be in public hands)	172.637
	Szeged	B, TB	Szegedi Közlekedési Korlátolt Felelősségű Társaság Tisza Volán	Public (it is used to be partly private before bankruptcy) Private (for 51% share in transport it gets 2/3 revenues)	164.883
	Pecs	B	Tüke Busz Zrt.	Public	156.649
	Debrecen	B, T, R	DKV Debreceni Közlekedési Zártkörűen Működő Részvénytársaság	Public	204.124
Slovenia	Ljubljana	B	JP Ljubljanski potniški promet, d.o.o.	Public	282.944
	Maribor	B	JP za mestni potniški promet Marprom, d.o.o.	Public	111.115
	Koper, Piran, Novo Mesto	B	Arriva Dolenjska in Primorska, druzba za prevoz potnikov, d.o.o.	Public (but international)	107.756
Croatia	Zagreb	B, T	Zagrebački električni tramvaj (ZET) HŽ (suburban railway)	Public (Zagreb holding) Public	790.197
	Osijek	B, T	Gradski prijevoz putnika (GPP) d.o.o.	Public	108.048
	Rijeka	B	KD Autotrolej d.o.o.	Public	128.624
	Split	B	Promet d.o.o.	Public	178.102
Bosnia	Sarajevo	B, T, TB	KJKP Gras d.o.o.	Public	275.524
	Banja Luka	B	Autoprevoz A.D.	Private	199.191
	Tuzla	B	Gradski i prigradski saobrać aj d.d.	Public	120.441
	Mostar	B	Mostar bus d.o.o za javni gradski prijevoz	Public	113.169
Serbia	Beograd	B, T, TB, R (under construction)	GSP Beograd Arriva Litas	Public Public (but international)	1.344.814

Country	City	Modes of transport	Urban transport operator	(Majority) ownership of the operator	No. of city citizens as per latest census
	Niš	B	JKP Direkcija za javni prevoz Grada Niša	Integral part of municipality	260.237
	Kragujevac	B	Gradska agencija za saobraćaj	Public agency	150.835
	Novi Sad	B	JGSP Novi Sad	Public	250.439
	Subotica	B	JP Subotica Trans	Public	141.554
	Mladenovac, Aranđelovac, Kragujevac, Obrenovac, Smederevo, S. Palanka, Inđija, Valjevo	B	SP Lasta, a.d.	Majority public	N/A
Montenegro	Podgorica	B	"Gradski saobraćaj PG" Podgorica d.o.o. "Bulatović trgopolje" d.o.o. "Montenegro prevoz Pejović" d.o.o.	Public Private Private	150.977
Macedonia	Skopje	B	JSP Skopje	Public	510.000

Source: Author's collection.

Table 1. Contracted public transport operators in selected cities with majority ownership and modes of transport under management.

The least frequent public transport mode is metro, available only in Prague, Budapest and Belgrade. These are the cities with the largest number of inhabitants and suburban area. The 142.4 km long tram network in Prague is one of the most developed in Europe with about 950 trams in operation. The main operator provides about 4/5 of bus connections, while the rest are rendered by private operators. Bus fleet is also very large with 1.255 buses operating on 148, 1.6783 km long lines. The metro network has three lines over 65.2 km, while the fourth line has been under construction. Other cities have much smaller public transport networks. Maribor with slightly over 100,000 inhabitants is served by about 50 buses only. Useful data on public transport organisation in smaller cities can be found in PROCEED country reports [10] or from CiViTAS project participants [11]. In smaller cities public transport is comprised of buses only that can be sometimes substituted by public bicycles. Although there are counters of bicycles in many areas, the statistics on bicycle usage for commuters changing buses for bicycles at certain points of their trip is not available.

Organisational structure of public city operators differs according to the number and complexity of public transport and related services. Transport operators are commonly responsible for various segments of public transport. There are urban areas that are serviced by several transportation modes such as buses, trams, trolleybuses, light rail, trains and metro. If urban transport operator is organised as a separate company, these organisational segments are often

http://dx.doi.org/10.5772/65751

not separated clearly as they use the same support services such as sale ticket offices, managerial and administrative staff support, public procurement, advertising, etc. It is especially hard to separate physical transport network construction and maintenance costs from the transport fleet costs. Sometimes a couple of public enterprises agree on the provision of certain service as it is the case with intermodal transport operated by various companies. If bus network is operated by one company and the rail network by another, then they often agree to provide the unique fare for regular passengers travelling from suburban to urban areas for the reasons of work and education, i.e. daily commuters. The citizens prefer efficient public services, such as buying fares in one company only. The urban transport operators share revenues and costs arising from the commonly sold fares thereafter according to certain allocation keys. In addition, there are many companies that are responsible for urban and suburban, as well as for regional, national and even international passenger transport (Lasta, Autoprevoz). Such companies can be registered for occasional passenger transport, international passenger transport, sale of spare parts for vehicles, as tour operators or other tourism supporting services. Public transport operators can also offer other services such as central bus station or metro station management, operating stations for technical inspection of buses, provision of advertising places in the public transport network, etc.

Municipal and public transport services often intertwine. Some public transport services such as road or rail infrastructure are commonly financed by local (sometimes central) authorities which often provide transfers to public transport operators for covering the capital costs of the public transport fleet purchase. In addition, municipalities strive to provide quality public service, thus combining the transport fares with other tickets for city sight-seeing spots, museums, libraries, sports facilities, parking lots, cable-car rides, even certain shops and restaurants (Urbana card in Ljubljana, tourist cards in other cities). Most cities purchase new vehicles when coming to the fleet renewal. However, there are some examples of purchasing the second-hand urban transport fleet or using EU funds for both hard infrastructure and urban transport fleet renewal (Szeged). The town of Szeged is one of the rare cities that developed its own fleet by adapting and refurbishing the existing or second-hand fleet in the first decade of twenty-first millennium what was seen before in Ostrava. The benefits of such an approach are counted by [12], consisting mainly of spreading financing term and thus achieving better financing conditions, reallocation of the workforce from service to production work, cooperation with industrial partners, tailoring fleet according to their own needs, better communication with the citizens.

3. Operating performance of city public transport operators

The data on operating performance have been collected by means of the European business registry and the websites of public transport operators. Financial and ownership data have been obtained via Amadeus database, public authority or public transport operator websites. Urban population data have been obtained from the most recent census data available. In the EU countries, the last census dates from 2011, while non-EU countries had it either in the same year (Serbia) or within the next two years (Bosnia and Herzegovina).

Only largest companies disclose their nonfinancial and financial data. Most transparent urban transport operators are Dopravní podnik HL.M. Prahy (Prague) and some local public operators in Serbia and Croatia that are obliged by law to make their annual financial reports public. Due to small number of large cities and old and inconsistent financial data available for different years, Slovakia is represented by only one town—Košice, while the data were insufficient for analysing public transport in Montenegro, Macedonia and Albania.

Even when disclosed, financial data on public city operators are often not comparable. It is explained by different modes of transport (tram, bus, metro and trolleybus), number and age of vehicles, number of urban and suburban lines and areas covered, number of passengers transferred/passenger kilometres passed, passenger structure, pricing policies, frequency of public transport usage by local citizens, different cost structure of public transport operators, etc. In general, public transport operators disclose a couple of general data about themselves: the transport network map with zones, the daily and night timetable and the price of typical fares.

Table 2 provides the data on city operators' indebtedness, fleet value per citizen (tangible assets), sales earned per citizen and per employee, number of citizens served by operator's employees and operating performance per employee. Contrary to expectations, public transport operators are not indebted although many of them end the year in loss. It means that the public transport fleet purchase is primarily financed from the local budgets or other grants. The only city transport operators that have some portion of long-term debt in assets are those of Koper (Arriva), Ljubljana public transport operator and Autotrolej in Rijeka. Since assets are presented net of depreciation, higher assets per citizen imply higher public transport fleet costs/newer fleet. The differences between the city public transport operators according to tangible fixed asset criterion per citizen are very large. It ranges from 1.17 euro per citizen in Maribor to 1.96024 euro in Prague. Average tangible fixed assets in analysed cities are 273.36 euro per citizen. When this indicator is coupled with the fare revenue earned per citizen, it is evident that the cities in which public transport service is used most are Prague, Budapest, Ljubljana and Brno. Publicly owned transport operator in Ostrava has significant asset value per citizen, but revenues earned per citizen are rather small, suggesting inadequate pricing policy of the operator/ public authority or small usage of public transport by the citizens. Just opposite holds for Ljubljana, where fare revenue is significant compared to the tangible fixed asset value per citizen. Other public operators, except for Belgrade, are smaller in size. Public transport is to some extent used in Croatian towns of Split and Rijeka, Plzeň, Miskolc and Debrecen, Novi Sad, Belgrade, Sarajevo and Ostrava (Arriva). All public operators having the value of tangible fixed assets per citizen lower than 100 euro will soon have to replace their public transport fleet unless the fleet is too small compared to the number of citizens that use public transport. The citizens of some cities are not used to public transport after they leave school (Mostar), while some cities purchase public transport fleet on operating leasing contract (which might be the case in Maribor). If sales per tangible fixed asset ratio is high, it suggests that the urban transport fleet may be close to the end of its depreciation

Country	City	L-T debt/ tangible fixed assets	Tangible fixed assets per citizen	Number of employees per 1000 citizens	Sales per number of citizens	Sales per employee	EBITDA per employ- ee	EBIT per employee
Czechia	Praha	0.38%	1.96024	7.76	444.00	57.24147	11.86466	1.32381
Hungary	Budapest	1.55%	1.17650	6.83	247.47	36.21312	7.21849	728.77
Serbia	Belgrade	4.96%	164.17	4.42	87.35	19.76344	1.08209	−950.05
Slovenia	Ljubljana	30.60%	112.65	3.08	160.58	52.11521	5.75182	395.91
Czechia	Brno	0.00%	505.62	7.13	102.46	14.37882	9.26154	2.84076
Czechia	Plzeň	1.59%	365.27	3.99	85.15	21.34820	12.55950	3.10179
Czechia	Ostrava (Arriva)	0.00%	82.15	3.83	70.41	18.36300	6.65116	1.19047
Czechia	Ostrava	0.00%	436.89	6.90	66.13	9.58139	4.65395	−23.12
Czechia	České Budějo- vice	1.82%	185.75	2.43	40.79	16.81539	7.95029	536.23
Slovakia	Košice	5.73%	159.75	6.24	57.96	9.29072	1.02100	−851.98
B&H	Sarajevo	3.98%	185.40	6.08	70.36	11.56691	−5.02872	−6.72550
Croatia	Split	10.31%	107.08	4.47	85.88	19.21495	2.32425	531.65
Hungary	Szeged	5.49%	219.86	3.20	43.19	13.51315	6.75275	1.07305
Croatia	Rijeka	28.75%	53.19	4.76	84.81	17.82362	3.53499	379.54
Croatia	Osijek	18.74%	279.14	3.24	39.53	12.20277	3.40359	579.90
B&H	Tuzla	9.93%	29.50	1.86	42.41	22.80271	2.24603	232.82
B&H	Mostar	11.84%	3.17	0.75	14.58	19.41706	336.85	162.41
Czechia	Hradec Králové	0.00%	228.49	2.58	43.04	16.68437	5.83218	36.55
Czechia	Olomouc	0.00%	141.00	2.32	37.26	16.05940	−3.92207	−9.13162
Hungary	Pecs	22.32%	22.78	2.96	64.38	21.73573	380.65	−29.52
Hungary	Debrecen	0.00%	245.70	3.32	82.77	24.95623	1.88297	−239.16
Hungary	Miskolc	9.05%	477.39	4.62	93.35	20.22125	2.38660	−1.60400
Serbia	Subotica	0.00%	42.57	3.35	56.42	16.84331	1.27759	107.88
Serbia	Novi Sad	6.10%	67.13	5.12	96.81	18.91164	1.08680	−141.48
Slovenia	Maribor	0.00%	1.17	1.48	28.50	19.19590	−72.49	−252.37
B&H	Banja Luka	11.04%	55.73	1.21	19.66	16.24834	−2.52192	−3.67060
Slovenia	Koper	40.44%	72.41	1.98	73.13	36.99545	8.92766	3.72974
	Mean	8.32%	273.36	3.92	86.61	21.46309	3.58675	−246.97

EBITDA (earnings before interest, taxes, depreciation and amortisation); EBIT (earnings before interest and taxes), L-T debt (long-term debt).

Table 2. Selected indicators of public transport operator assets, sales, indebtedness and operating results.

period or that the value of public transport fleet is very small compared to the size of the city (Tuzla, Rijeka, Pecs, Subotica, Novi Sad and Koper). Sales per number of citizens reveal the importance of city public transport in certain cities. The highest fare revenue per citizen is earned in the city of Prague, Budapest, Ljubljana and Brno, suggesting that public transport is vital in the largest urban areas. Organisational issue of city public transport is well revealed by the fare revenue, i.e. sales per employee. Well-organised public transport according to this criterion exists in Prague (57.2 thousand euro per employee) and Ljubljana (52.1 thousand euro per employee). Quite well efficient are public transport operators in Budapest and Koper, while least revenue per employee (less than 15 thousand euro) was earned in Sarajevo, Osijek, Ostrava, Košice, Szeged and Brno in 2013.

In most cities fare revenue is the main source of operating revenues. For operators in Bosnia and Herzegovina, operating revenue data was missing; thus, it was estimated that it is equal to fare revenues. However, there are cities is which ticket revenues are well below 50% of the operating revenue threshold. It suggests that public transport operators are heavily subsidised by public authorities. Such cities may be Osijek and all Czech cities except for Prague, Szeged, Koper and Maribor.

Public transport operators typically collect fares on their own. Although it appears simple when public transport is operated by one operator, it can be very complicated when there are more public transport operators in place. Thus, many cities opted for gross contract principle in which operational (revenue) risk rests with the public authority.

Public transport operators should at least cover their operating costs, i.e. their earnings before depreciation, interest and tax should be positive. The average data show that operating result before depreciation is positive, while adding up depreciation turns the operating result per employee into negative area. The most efficient cities measured by operating result accomplished per employee are Brno, Plzeň, Ostrava (Arriva), Koper and Szeged. However, Olomouc, Sarajevo, Miskolc, Belgrade and Košice have substantial negative results. The average number of employees in public transport operators per citizen is four per 1000 citizens. The cities with the highest ratio include Praha, Brno, Budapest, Ostrava (public operator), Košice and Sarajevo.

However, many public transport operators do not even have substantial fare revenues to even cover their staff costs. This happens in Sarajevo, Osijek, Brno, České Budějovice, Košice and Maribor. These cities, despite positive earnings before depreciation, are either transferring certain funds to public transport operators for the salaries or for the fleet or are simply giving them subsidies to earn a positive operating result. In all analysed cities, the portion of sales used to cover staff costs is higher than for covering material costs. In the end of 2013, the loss per employee was the highest in Sarajevo at more than 8000 euro, which was also significant per citizen (over 49 euro). Cities of Olomouc, Miskolc, Prague and Belgrade have sustainable loss per citizen ranging from 10 to 18 euro. Overall, 17 out of 27 public transport operators finished the year in loss, while 10 of them showed positive results.

http://dx.doi.org/10.5772/65751

4. Financing public transport operators

Revenues to public transport operators come from passengers paying either full, discounted or partly subsidised fares. Full fares are calculated for one-time travellers; return fares and fares valid a month or a couple of days are commonly sold at the discount either to frequent passengers or to tourists. In largest cities and popular tourist destinations, public transport fares are bundled to tourist tickets for sightseeing. They can be offered for 1, 2, 3 or 7 days. The longer the stay, the cheaper the ticket. In some cities the fares are more expensive when purchased in the vehicle than when purchased in designated shops or kiosks on stations, while some cities calculate the same price for the fare regardless of the place of its purchase (Zagreb). Fare price are in most cities the same for daily or night ride, but some cities charge more expensive fares for night ride (Košice, Zagreb) [13]. However, night lines are less frequent and on many routes non-existent. Partly subsidised fares are valid for certain groups of citizens such as school children, students, disabled people, the unemployed or the retired. The rest of the costs up to the full fare price is covered from the local or national budget. There are also partly subsidised fares for other end users such as workers of certain companies which are at the level of employer as the subsidy is paid from the corporate entities either to the user of public transport service or to the public transport operator. Monthly or annual fares can be purchased by a certain person, and they are not transferable. There are, however, time tickets, quantity tickets and value cards which are transferrable from person to person. Time tickets are typically valid from 1 to 7 days, while quantity tickets are issued for 2–30 rides. Value cards are prepaid cards with a certain amount of (prepaid) credit that can be used for purchasing the fares. A single fare is typically valid for 60- or 90-minute ride throughout the selected number of city zones. The fares are usually the same for different city transport modes operated by the same company (tram, trolleybus, bus, city metro or city rail). The city of Košice has the fares for 30- and 60-minute ride, and it even has a reduced fare of 0.25 euro for up to four stops of single ride, while Bratislava distinguishes between 15 and 180 minutes of rides at prices varying from 0.7 to 3.6 euro [14]. České Budějovice has 20- or 60-minute ride option, while Ostrava even introduced 10-minute ride ticket. Paper tickets still exist in many cities, while some cities are stimulating electronic purchase. If purchased by SMS, the ticket price is cheaper in Košice for 20 cents, for instance. The Slovakian cities have extra charge for luggage and pets [13]. Some public transport operators stimulate the demand for public transport services by enabling variable fees in non-rush hour time and on weekends (Subotica). Miskolc has introduced weekend family ticket of approximately 9.35 euro. An income census is sometimes introduced in determining fare prices. Such a system exists in Sarajevo and Osijek for senior citizens. Some cities have a gradation between pensioners' age, whereby the oldest citizens pay the cheapest passes. The fares charged to pensioners can be the same as those priced for students (Pecs, Belgrade, Ljubljana, Koper, Piran and Novo Mesto), lower than student passes (Hradec Královė, Osijek, Rijeka, Budapest, Subotica, Tuzla, Banja Luka, Mostar and Sarajevo) or higher than student passes (Split, Brno, Prague, Novi Sad and Maribor). Sarajevo public transport operator also introduced school children and students passes based on the distance travelled, in which students travelling up to 2 km distance subsidise those travelling more than 2 km. Even though Sarajevo public transport operator strived to be fair in pricing, it comes at the cost of a very complicated public transport pricing policy with many grades of fare prices that is very difficult to control. Cross-subsidies to senior citizens are also very obvious in Osijek, Zagreb, Mostar and Hradec Králové.

Different daily and monthly fare prices among cities are shown in graphs 1–3. **Figure 1** shows single-fare prices bought at the kiosk. The fares bought in the vehicle are usually 10–30% more expensive. The highest single-fare prices have been observed in Croatian cities and Košice. The city of Split has the same single-ride price as Ljubljana, Belgrade and Bratislava, while Prague and Budapest have somewhat lower prices. The average single-fare price is 0.96 euro, whereby all capital cities (except for Sarajevo) and other Croatian cities, Brno, Miskolc and Košice, have higher prices, with ten cities charging lower prices than the average. The lowest single-fare prices are valid in Novi Sad, Subotica and Tuzla.

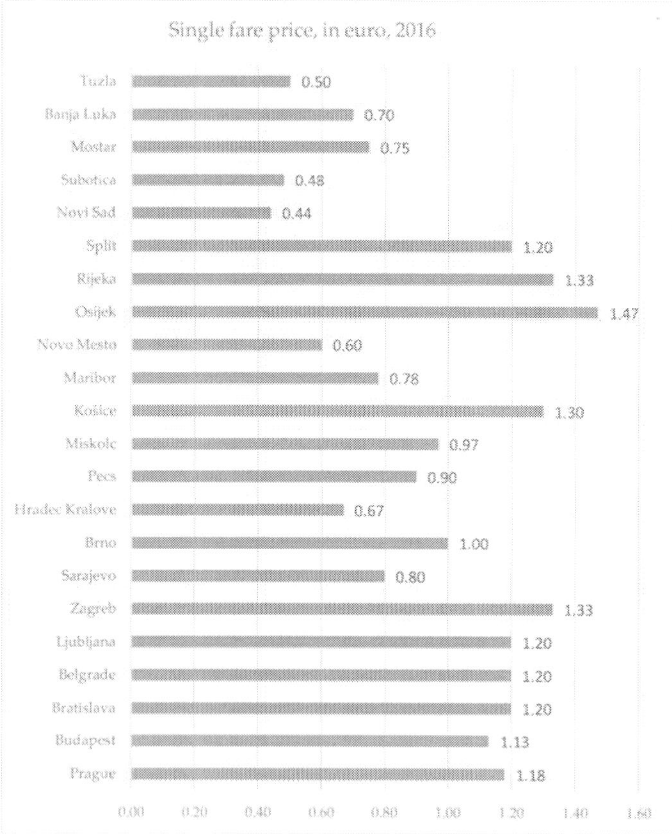

Figure 1. Single-fare prices in selected cities, in euro, 2016. Source: Author's collection.

http://dx.doi.org/10.5772/65751

As illustrated in **Figure 2**, Zagreb has the highest prices of monthly adult passes at 48 euro, followed by Ljubljana (37 euro) and Budapest with 33.87 euro. The average monthly pass price for an adult in capital cities is 29.61 euro. Contrary to expectations, Prague has the cheapest public transport with monthly pass price of only 11.10 euro. This fact sheds some light on the losses made by Prague transport operator, which are not at all too high when fare prices are taken into calculation. The profit and loss account data show that each citizen of Prague has to give up about 10 euro for making up for the losses in public transport. That is actually a price increase of public transport by less than one euro monthly. Even with this increase in price, Prague public transport would still be well below the comparable prices in other cities.

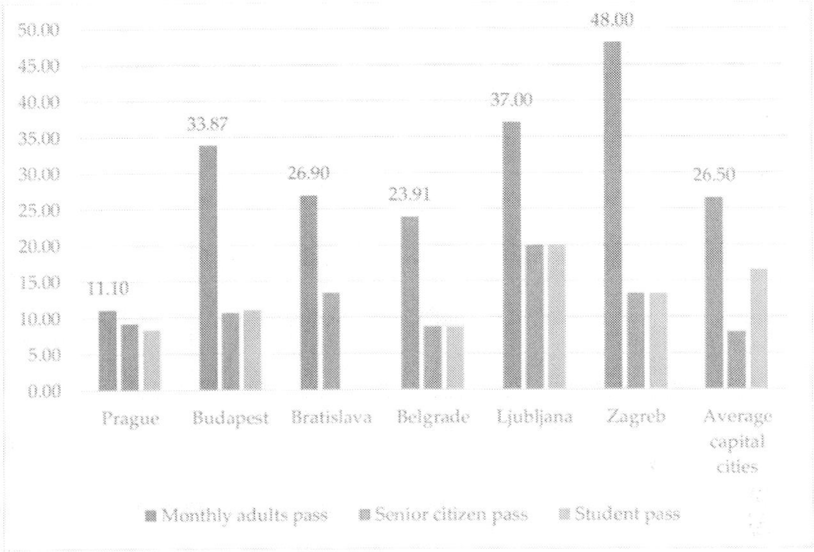

Figure 2. Monthly prices of fares in capital cities, in euro, 2016. Source: Author's collection.

Figure 3 shows the oscillations around the average price of monthly adult pass among smaller public transport operators of 24.02 euro. Hereby, Croatian cities and Brno are at the levels higher than 30 euro. It can be easily spotted from **Figure 3** which city has lots of pensioners, where pensioners are assumed to have a decent living standard (Brno and Novo Mesto) and where pensioners rather belong to a social category of citizens (Osijek, Banja Luka and Hradec Králové). The average difference between adult and senior fare monthly pass is significant, ranging from 18.5 euro in capital cities to 6.19 euro in non-capital cities. Such huge difference warns of heavy cross-subsidisation of public transport in favour of pensioners in capital cities. However, the discounts offered to students are much smaller, ranging from 10 euro in capital cities to only 2.36 euro in non-capital cities on average. Herby, Croatian non-capital cities and Novi Sad have over 50% price reduction for monthly student passes compared to adult passes, while the largest reduction for student passes exists in Zagreb (over 72% of the adults pass).

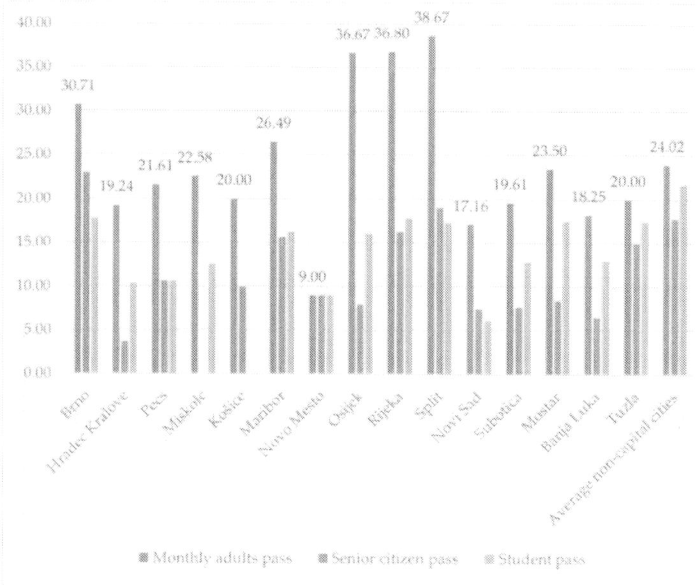

Figure 3. Monthly pass prices in selected non-capital cities, in euro, 2016.

The financial inflows and outflows for public transport are shown in **Figure 4**. Apart from fares, other revenues of public transport operators come from local or national taxes whose portion is allocated for transport infrastructure, special taxes for funding transport infrastructure, local administrative and other fees (cross)-subsidies/grants and loans. Subsidies and cross-subsidies are more specific as they may and may not come from the direct beneficiaries of the transport infrastructure. Subsidies occur when the funds from local budget are transferred directly to public transport operators, usually for covering their deficits. Cross-subsidies occur when revenues in excess of variable costs for one group of passengers are used to finance deficits incurred for other groups of passengers [15]. This situation is common for financing fares of school children, students, the disabled or senior citizens. Cross-subsidies can also occur when funds from other profitable activities, but public transport, of either local municipality or public transport operator are used to make up for the difference (deficit) in public transport operations. There are cases when taxes imposed to the direct users of transport infrastructure serve as general rather than tax revenues for transport infrastructure financing, which often happens in highly centralised and highly indebted countries. Funds are commonly available for transport systems funding from international financial institutions either in form of one-off grants or in form of the loans with more favourable interest rate or various kinds of technical assistance for public transport system planning, operation and maintenance. In all these cases, funds come from indirect beneficiaries of transport infrastructure. International institutions find their interest in connecting the regions and people for the purpose of equal development and easier transfer of goods, services and people.

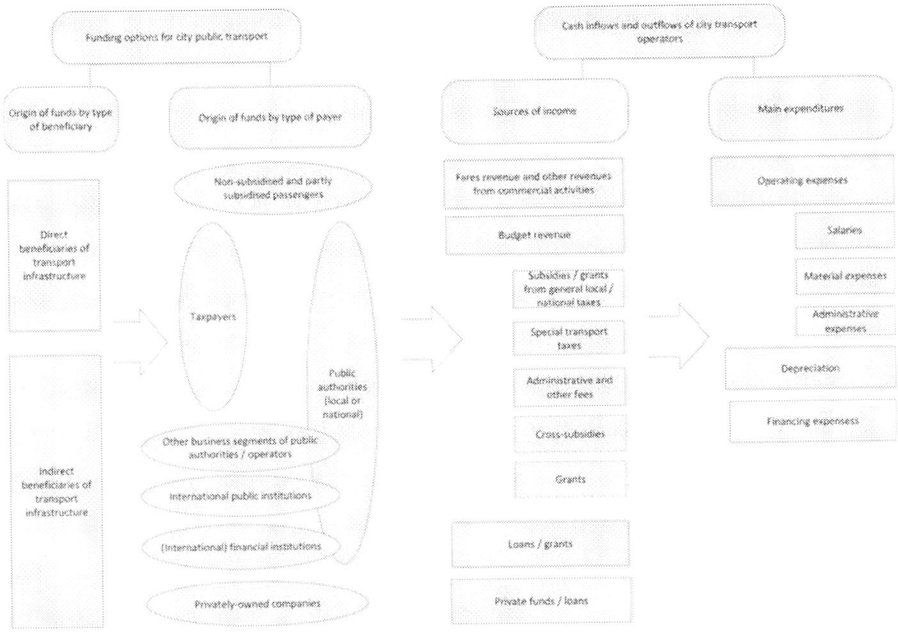

Figure 4. Urban transport operators financing mechanism. Source: Author's illustration.

Most public transport operators do not have sufficient commercial revenues to operate the system in a sustainable way. Local and sometimes national governments provide funds for loss coverage. The size of loss is directly linked to the size of the city and suburbs' population, fleet size, quality and age, fuel consumption and pricing policy. The costs per vehicle vary depending on the type of vehicle (tram, train and bus), fuel type, producer, size, carriage, number of seats and number of standing places and special characteristics such as access to disabled people, spare parts, service costs, etc. These costs are generally disclosed for the urban transport fleet when financed from the loans/grants of international financial institutions or on news portals in national languages only. The public transport operators are more ready to disclose the data on emissions and type of vehicles they have in their fleet, which is beneficial for sharing technological know-how in public procurement [16]. Besides technology, spatial and institutional problems the cities are faced with, financial problems with public transport renewal are most frequently cited among the cities. Thus, financial affordability of the cities influences very much the choice of public transport technology and hence the quality of public transport fleet. Not many Central European cities have prepared projects for urban transport fleet renewal during the 2007–2013 period, while Southern European countries have not done it at all. An exemption is Bratislava that purchased 15 monodirectional and 15 bidirectional trams for 220 passengers each, for 91.26 million euro, of which 61.41 million was co-funded by the European Regional Development Fund (ERDF) [17]. Szeged also prepared a good plan for urban transport renewal which consisted mainly of physical infrastructure renewal that included new and old vehicle refurbishment. However, European cities benefited significantly for transport infrastructure renewal and construction, mostly for cross national, national and regional, but also for urban transport (metro line extension in Prague, metro, tram and suburban railway infrastructure in Budapest). If the private sector is involved in public transport fleet renewal, it generally assumes the licence to hold a certain route(s) for a specified number of years. However, there are examples of urban transport fleet financed entirely by the private sector at no cost for the citizens. The latter occurs in Plzeň for route towards a shopping centre of which the shopping centre has direct benefits [12].

Capital costs of urban fleet, though high, are smaller than its operational costs during the fleet expected lifetime. Public transport fleet can be funded from a number of sources. Litman [18] counts 18 sources of funding city public transport: 15 belong to direct or indirect taxes related to either fuel consumption, gas emissions, road or property taxation, full price and discounted fare revenues and advertising revenue. In practice there are much more of them.

Determining a sustainable pricing policy for public transport operators is very challenging as it includes both commercial and social (politically sensitive) component. Real cost of service is for this reason often not disclosed or hidden. It can be explained by various factors such as company structure of city transport operators, organisational structure of city transport operators, intertwining of municipal and public transport operators' services, deliberate data intransparency, etc. Very few cities such as Ljubljana, Belgrade and Prague disclose some general information on public city transport like average age of public transport fleet, structure of fleet, urban and suburban area covered by public city lines, etc. However, even such operators may not update such information regularly.

Deliberate data intransparency assumes hiding the data from the public due to numerous reasons such as subsidies from municipal authorities and other public institutions, underperforming operating indicators due to high number of the employed, obsolete public transport fleet, huge material costs, hiding the data from the competitors (in the areas where there are several concessionaires and/or a public and private-operated public transport service), etc. Undeliberate data transparency is common for many municipal services. One explanation may be that public transport functioning is taken for granted by the citizens who do not need to get bothered by statistical data. The news on purchase of new vehicles is regularly provided by public transport operators, while other data are scarcely found. Luckily, the public transport operators are obligors of the public procurement which forces them to take account of the public transport fleet costs. The data on the costs of urban transport fleet replacement for some public transport operators are shown in **Table 3**.

Private operators are obliged to keep costs under control as they are requested to renew the contract with new fleet or fleet in good condition. They sometimes do it at the cost of higher operating costs as gross cost contracts do not stimulate them for cost reduction. Practically, all cities, regardless of the ownership structure of the operator, cover the difference between the operating costs set in the contract and fare box revenues.

City	Timing	Type of fleet renewal	No	Total price in million euro	Average price	City budget participation	Way of financing
Belgrade	2009–2013	Trams	30	81.3	2.71	75%	
		Trolley buses	83	18.6	0.22	100%	
			16	3.8	0.24	100%	
		Fare collection and fleet management	-	12.5		N/A	
Zagreb	2008	Trams	70	N/A	N/A	100%	Leasing
	2009	buses	214	N/A	N/A	100%	Leasing/later sale and lease back to public authority
Bratislava	2007–2013	Trams	45	115.35	2.56	20%	Budget/ERDF
Prague	2015	Trams	34	81.4	2.39	N/A	EU funds/ municipality
Szeged	2004	Second-hand trams	14	4.6	0.33	100%	Bank loan
	2003–2004	Refurbishment of the trams	10		0.00		
		Trolleybus	8		0.00		
	2011–2012	Trams	9		0.00		EU funds/ municipality
		Articulated buses	10		0.00		
Košice	2010	CNG buses	19	9.12	0.48	100%	
	2013–2014	Trams	23	48	2.09	100%	
		Buses	127	28.6	0.23	100%	
Osijek	2009	Buses	12	N/A	N/A	100%	Financial leasing
Rijeka	2013	Buses	11	3.33	0.30	100%	Loan
		Minibuses	10		0.00		

Table 3. Collected data on urban transport fleet financing.

5. Conclusion

Comparing public transport operation financing in different cities of the transition economies of Central and Eastern Europe and Southern and Eastern Europe turns out in many aspects to comparing the incomparable. Not only public transport operators differ in size, organisation, ownership, number of employees and fleet characteristics (various modes of transport, varying condition of infrastructure, different public transport fleet quality measured by manufacturers, age, capacity, energy consumption and congestion, accessibility, comfortability) but lots of behavioural and political elements are always involved in city public transport pricing and functioning. Citizens may be reliant on urban public transport to lesser or greater extent, which very much depends on the city organisation, average time necessary for transfer over a particular route by public or private transport and price of public transport. There is no unique statistics on city public transport functioning, so the data shown very much depend on city public transport operators, local authorities, countries, case studies done within the scope of certain international transport projects (PROCEED and CiViTAS) or international studies. Not only city public transport data are coupled with the regional and national numbers in national statistics but some basic data are not available at all on regular basis or are available for local citizens only. The data collected from various sources suggest that urban transport statistics should be more comprehensive and the data unified across the countries. Disclosing the data on approximate time of public transport in one direction, the number of passenger kilometres passed by transport modes, the revenue by type of passengers, operating costs, the amount of revenues, subsidy size, number of employees and fleet book assets and age would contribute very much to the international comparison of public transport operators' efficiency.

Public transport operations are in a nutshell always financed by the citizens. However, they may be financed either directly through fare revenue (which is the fairest approach) or indirectly through various taxes collected from both those who use it regularly and from those citizens who do not use it at all. The portion of direct and indirect user financing is dependent on the decision of the local public authorities. The typical way of financing public transport operators is to cover at least operational costs by fare revenues, while budgets make up for the rest. Only regular disclosure of the data on public transport funding in local budget would reveal the level of public transport sustainability. Larger cities, although faced with the much greater complexity of public transport organisation, have a privilege of disposing with much larger budget that can hide public transport operator inefficiencies. Even if the funds are not sufficient, rare public transport operators end up in loans. Rather, the local authority borrows funds to keep the quality of either transport fleet or physical infrastructure. The international financial institutions can help very much in covering occasional costs related to public transport fleet renewal. Although many cities opt for applying for transport network co-funding, the projects can combine investments in fleet renewal. The competition of private sector is very limited in the analysed countries, but it gradually emerges. All public transport operators, regardless of their ownership structure, generally operate on a gross contract basis, whereby the city makes up for the losses. However, only few cities have introduced penalty-reward mechanism to encourage the efficiency of private partners. Overall, there is some awareness on growing importance of energy efficiency of public transport evidenced in purchase of quality transport fleet, but there is a lot to do in integrating public transport modes (including the bicycles) and public transport promotion, especially in the countries of the Southern and Eastern Europe.

http://dx.doi.org/10.5772/65751

Author details

Mihaela Grubišić Šeba

Address all correspondence to: mihaela.g.seba@gmail.com

The Institute of Economics, Zagreb, Croatia

References

[1] UITP - De L'Union Internationale Des Transports Publics. Local public transport trends in the European Union. Statistics Brief. June 2014. [Internet]. Available from: http://www.uitp.org/sites/default/files/cck-focus-papers-files/Local_PT_in_the_EU_web %20%282%29.pdf.

[2] Rodrigue J.-P., Comtois, C., Slack, B. The Geography of Transport Systems. 1st ed. Routledge – London and New York; 2006.

[3] European Commission. White paper – Roadmap to a single European transport area – Towards a competitive and resource efficient transport system [Internet]. 2011. Available from: http://eur-lex.europa.eu/legal-content/EN/TXT/PDF/?uri=CELEX:52011DC0144&from=EN [Accessed: 10/07/2016]

[4] Posada F., Chambliss S., Blumberg K. Costs of emission reduction technologies for heavy-duty diesel vehicles [Internet]. 2016. Available from: http://www.theicct.org/sites/default/files/publications/ICCT_costs-emission-reduction-tech-HDV_20160229.pdf [Accessed: 10/07/2016]

[5] MJB & A Strategic Environmental Consulting. Clean diesel versus CNG buses: Cost, air quality & climate impacts [Internet]. 2012. Available from: http://www.catf.us/resources/publications/files/20120227-Diesel_vs_CNG_FINAL_MJBA.pdf [Accessed: 31/07/2016]

[6] Unknown. Maintenance costs: Comparative analysis between diesel vs CNG fuelled vehicles [Internet]. 2014. Available from: http://www.theammj.com/1topics947roch-coe67/t274/14diesel.pdf [Accessed: 31/07/2016]

[7] European Commission. Statistical pocketbook 2014 – Transport [Internet]. 2015. Available from: http://ec.europa.eu/transport/facts-fundings/statistics/pocket-book-2014_en.htm [Accessed: 31/07/2016]

[8] Trolley Project Central Europe (co-funded by ERDF). Trolley transport mode efficiency analysis – Comparison of financial and economic efficiency between bus and trolleybus systems [Internet]. 2013. Available from: http://trolley-project.eu/

fileadmin/user_upload/download/TROLLEY_WP4_Transport_Mode_Efficiency_Analysis_Bus_vs_Trolleybus.pdf [Accessed: 31/07/2016]

[9] UITP. Metro, light rail and tram systems in Europe [Internet]. 2010 . Available from: http://www.uitp.org/sites/default/files/cck-focus-papers-files/errac_metrolr_tramsystemsineurope.pdf [Accessed: 31/07/2016]

[10] Various countries for EC, DG for Energy and Transport. National reports of PROCEED (Principles of successful high quality public transport operation and development) [Internet]. 2007. Available from: http://www.proceedproject.eu/index.php?option=com_content&task=view&id=177&Itemid=29 [Accessed: 31/07/2016]

[11] CiViTAS Initiative. Cities directory [Internet]. Available from: http://www.civitas.eu/cities-directory [Accessed: 31/07/2016]

[12] Bartłomiejczyk, M., Polom M. (eds).Determinants of Functioning of Trolleybus Transport in Selected Cities of the European Union. Bernardium, Pelplin. 2011.

[13] IMHD.SK. Public transport in Košice [Internet]. 2013. Available from: https://imhd.sk/ke/doc/en/11664/Tickets-from-1st-Jan-2013 [Accessed: 31/06/2016]

[14] IMHD.SK. Ticket price list – Public transport in Bratislava [Internet]. 2015. Available from: https://imhd.sk/ba/doc/en/14971/Ticket-Price-List-from-1st-Nov-2015 [Accessed: 31/07/2016]

[15] Ubbels B., Nijkamp P., Verhoef E., Potter S., Enoch M.. Alternative ways of funding public transport – A case study assessment. EJTIR. 2001;1(1):73-89.

[16] CiViTAS POINTER. Cluster report 2: Clean Vehicles and Fuels [Internet]. 2013. Available from: http://www.civitas.eu/sites/default/files/d2.6.3-pointer-pu-cluster_report_2_clean_vehicles_and_fuels-14-nov-2013-final-trg_tno.pdf [Accessed: 31/07/2016]

[17] European Commission – InfoRegio. New trams for a more efficient transport system [Internet]. 2014. Available from: http://ec.europa.eu/regional_policy/en/projects/slovakia/new-trams-for-a-more-efficient-transport-system [Accessed: 31/07/2016]

[18] Litman T. Evaluating public transportation local funding options, J Public Transport. 2014;17(1):43–74

[19] Steer Davies Gleave. Study on economic and financial effects of the implementation of Regulation 1370/2007 on public passenger transport services [Internet]. 2016. Available from: http://ec.europa.eu/transport/themes/pso/studies/doc/2016-02-effects-implementation-regulation-1370-2007-public-pax-transport-services.pdf [Accessed: 31/07/2016]

Chapter 10

The Rise of Ride Sharing in Urban Transport: Threat or Opportunity?

Tamer Çetin

Additional information is available at the end of the chapter

http://dx.doi.org/10.5772/66918

Abstract

Recently, the ride-sharing services have become very popular in urban transport. In particular, the rise of dynamic ride sharing startups have challenged to the traditional taxicab services. Although those dynamic developments within urban transport systems have made ride share services substitute for cab services, it has been controversial in the political sphere. Some researchers and policy makers argue that ride-share will lead to the death of public transportation, if government does not intervene in this area. However, market developments suggest the opposite. While competition between those urban transport modes has led to the dramatic decline in taxi fares, consumer welfare and economic efficiency have improved. Examining the rise of ride sharing and its effect on the traditional taxicab market, this chapter concludes that ride share services introduces a unique opportunity for the users of urban transport, but not threat. However, a mixed regulatory structure including deregulation and regulation policies is needed to improve market welfare in both markets. While economic regulations such as fare controls and entry restrictions are not necessary for both ride shares and taxicabs, social regulations are still crucial to improve the satisfaction of users of rideshare and taxicab services in terms of the quality of service.

Keywords: urban transport, Uber, ride sharing, taxicabs, regulation, deregulation, market welfare, efficiency

1. Introduction

This chapter examines the rise of ride sharing in urban transportation and its effect on the traditional taxicab services. The aim is to evaluate whether ride sharing is threat or opportunity for the future of urban transportation. The rise of ride sharing has influenced the nature of urban transport in many different areas from the quality of service to the structure of prices and demand within public transportation. The Uber experience suggests that ride sharing in taxi markets has become substitution for the traditional cab services. Clearly, the ride-sharing system competes with the traditional taxicab service. The success of ride sharing stems from its impact on consumer welfare. This refers to economic efficiency considerations under competitive market conditions. Most importantly, this system eliminates the transaction costs of a taxicab market under regulated market conditions [1, 2]. Conversely, the failure of traditional taxicab services stems from distortions inherent in a regulatory process. Price controls and entry regulations lead to artificial monopoly rents through medallion prices and taxi fares. Clearly, while ride sharing increases consumer welfare, the strictly regulated taxicab markets lead to death-weight loses in consumer welfare and economic inefficiency [3].

For that reason, in order to better understand the effect of ride sharing as a transport mode in urban transportation on the traditional taxi services, we also have to take into consideration the reasons and results of taxicab regulation. For this aim, the chapter provides an overview of the literature on the regulation of taxis. This section of the chapter includes the economic rationale for regulation in the traditional taxicab markets, the lessons from some regulatory experiences, and the effect of fare controls and entry restrictions in practice. The chapter also discusses the impact of ride sharing on the cab service and its regulation in the traditional taxi markets. It concludes with a critical statement on regulation, deregulation, and competition in taxi markets, including some policy suggestions. In this context, the chapter consists of three main sections including introduction. In Section 2, I evaluate the rise of ride sharing and its effect on taxi services in urban transportation. In Section 3, I discuss the regulation of taxicabs and introduce some policy suggestions about regulation, competition, and deregulation by taking into account the rise of ride sharing.

2. The rise of ride sharing or invisible hand

A substantial recent development in urban transportation has been the occurrence of ride-sharing system as the rise of invisible hand throughout the world. This intra-city transport mode occurred in San Francisco as Uber and Lyft and has become very popular in the United States in a short time. While Uber entered into the market in San Francisco in 2010, Lyft launched in June 2012. Since 2010, ride sharing has expanded and gained significant market share throughout the world. As of September 2015, Uber performs in 60 countries and 300 cities. Today, Uber is in use in 507 cities throughout the world.[1] It has an estimated market value of over $68 billion.[2] In the case of many cities, ride sharing started to be substitute for the traditional taxicab services in the intra-city public transportation. One recent report found that ride sharing met an average 46% of all total paid car rides through Uber in the major US cities in the first quarter of 2015,[3] because ride sharing is cheaper than a cab trip.[4] Note that this happened within 5 years.

[1] See https://www.uber.com/.

[2] See for information about statistics http://www.businessofapps.com/uber-usage-statistics-and-revenue and http://bruegel.org/2016/02/uber-and-the-economic-impact-of-sharing-economy-platforms.

[3] See for more detail http://www.forbes.com/sites/andrewbender/2015/04/10/ubers-astounding-rise-overtaking-taxis-in-key-markets/#2091ea0722ef.

[4] A recent study shows that ride sharing is cheaper than taxicab in 21 largest cities in the United States. See for more detail http://www.businessinsider.com/uber-vs-taxi-pricing-by-city-2014-10?IR=T.

Even though the dynamic ride sharing emerged recently, the ride-share services have been operating for a long time in the US cities. However, the dynamic ride-sharing services are remarkably different from the conventional ride-sharing applications, and in this sense, the success of the current ride sharing stems from its technological and dynamic nature. Even though the traditional ride-sharing programs were initiated during the World War II to save fuel, this initiation was not successful. However, in the 1970s, it reemerged as solution to environmental hazards from transportation such as energy consumption, air pollution, and traffic congestion. But, aiming at commuters, this system did not involve payment to the driver, even though contribution was made for gas or tolls in some cases. Mainly, the logic of ride-sharing service is to enable commuters to ride together on a regular basis. Park-and-ride lots and high-occupancy vehicles lanes that bypass congestion points in traffic support this system [4, 5]. The first-generation ride-sharing services were not successful. However, Uber has become one of the fastest growing startups in the sharing economy throughout the world. This system has appealed to people for some reasons. First, the dynamic ride sharing commits a regular ride-sharing arrangement for people who are willing to share a ride occasionally. Second, it is particularly attractive for people who are comfortable with computers and cell phone messaging and who are generally mobile in the modern cities. Third, registration and screening by a ride-share service makes this service safe and secure for commuters. Fourth, while the probability to find a quick ride-sharing match has increased over time, wait and pick-up/drop-off times have shortened [4, 6].

Today, the dynamic Uber or ride sharing connects drivers offering rides and passengers seeking them online through an online app downloaded on smart phones. This app is user-friendly and allows passengers to find the nearest available cars in the ride-sharing system. The cars are private. In other words, ride-sharing companies such as Uber and Lyft do not need to have their own cars. The companies sign up private drivers who are willing to provide rides to paying passengers and pass the ride requests directly to them. The system itself determines the price of the ride and all transactions happen through the online system. In general, 70-80% of each fare goes to the driver and the company keeps the rest.[5] A distinctive feature of ride-sharing service is that it does not include fare controls and barriers to entry as in the traditional taxicab markets. Even though courts in many countries banned or restricted ride-share services because of unfair competition considerations, competitive pressure from ride sharing to the traditional taxi services changes the nature of urban transportation. As a result, the dynamic ride sharing has increasingly become attractive to people who want to use public transport modes in urban transportation. This change in the preferences of consumer using urban transportation has made the ride-share services substitute for the traditional taxicab services. In order to understand this substitution relationship among urban transport modes, we have to look at the change in those transport modes. As depicted in **Figure 1**, while Uber ride increases directly correlate with taxi ride decreases, car rentals are relatively consistent. It is clear that demand for the traditional taxicab service decreases, and concurrently, demand for Uber increases. Because this change in the urban transportation modes continues in favor of ride sharing, we can infer that Uber has increasingly become substitute for the traditional taxicab services.

[5] http://bruegel.org/2016/02/uber-and-the-economic-impact-of-sharing-economy-platforms/

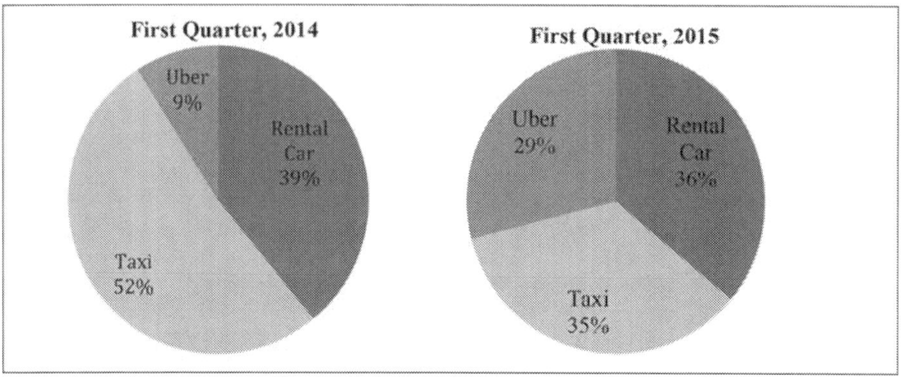

Figure 1. Uber vs. other urban transportation modes. Source: http://www.businessofapps.com/uber-usage-statistics-and-revenue/.

An important effect has been on the quality of taxi services. Wallsten [7] found that taxis respond to competition from the new ride-sharing services in New York City and Chicago by improving quality. The findings from this study confirm that the rise of ride sharing is associated with decrease in pre-trip complaints in New York. **Figure 2** clearly shows that the number of complaints has decreased after ride sharing entered the market in New York City. The findings from the case of Chicago [7] also suggest that complaints for the traditional taxi services have declined after Uber as ride-sharing service entered the market in Chicago. This research concludes that the results from New York City and Chicago are consistent with the idea that the traditional taxi services respond competition from the dynamic ride-sharing services by improving the quality of service.

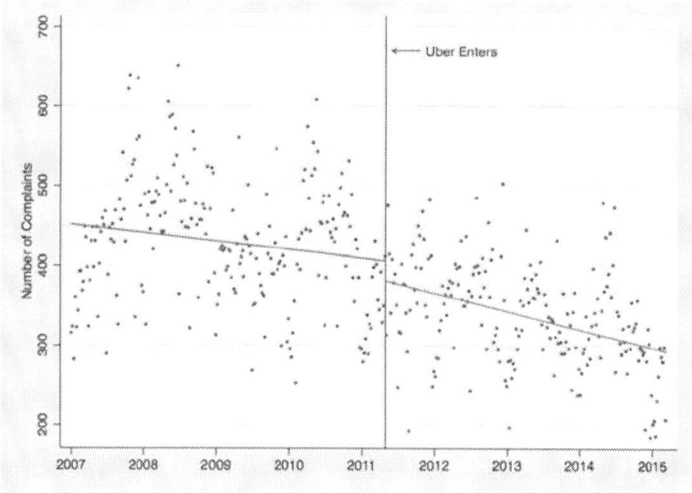

Figure 2. Number of taxi complaints submitted to New York City Taxi and Limousine Commission. Source: Ref. [7].

http://dx.doi.org/10.5772/66918

It is possible to state that the most important effect of ride sharing in urban transportation is the dramatic decline in prices. As reported in **Table 1** and **Table 2**, The rates for ride sharing are generally lower than regular taxis in most major cities in the United States, even excluding the taxi driver's tip. As summarized in Ref. [8],[6] "the column labeled Taxi/Uber shows the taxi fare relative to the Uber fare. If the ratio is over 1, as it is everywhere except New York and Philadelphia, that means that Uber is cheaper than a cab-that is, until surge pricing reaches that level. In L.A., an Uber car is cheaper for this sample trip even with surge pricing up to 1.7x. It is also important to note that you do not have to tip your Uber driver. And most people do tip their taxi driver. If you add a tip of 20% to the cab fares, Uber looks like an even better deal and beats out taxis in every city we analyzed." This finding clearly suggests that price competition leads to the expansion of ride sharing in urban transportation.

	Uber	Taxi	Taxi/Uber
New York	17.75	15.50	**0.9**
Philadelphia	15.25	14.20	**0.9**
Portland	15.05	15.00	**1.0**
Cleveland	13.00	13.95	**1.1**
Dallas	10.30	11.25	**1.1**
Miami	13.25	14.50	**1.1**
Indianapolis	11.65	13.00	**1.1**
Phoenix	11.00	12.50	**1.1**
Minneapolis	12.15	14.25	**1.2**
Baltimore	10.75	13.05	**1.2**
Columbus	10.20	12.85	**1.3**
Denver	10.35	13.75	**1.3**
Detroit	12.30	16.50	**1.3**
Seattle	11.70	16.00	**1.4**
San Francisco	12.30	17.25	**1.4**
Chicago	9.50	14.00	**1.5**
Boston	11.10	16.60	**1.5**
Atlanta	10.00	15.00	**1.5**
Houston	9.00	13.75	**1.5**
San Diego	11.35	17.80	**1.6**
Los Angeles	9.40	16.35	**1.7**

Source: http://www.businessinsider.com/uber-vs-taxi-pricing-by-city-2014-10?IR=T accessed on June 12, 2016

Table 1. Comparison between Uber and taxi fares.

[6] http://www.businessinsider.com/uber-vs-taxi-pricing-by-city-2014-10?IR=T accessed on September 2016.

	Uber	Taxi +20% Tip	Taxi/Uber
New York	17.75	18.60	**1.0**
Philadelphia	15.25	17.04	**1.1**
Portland	15.05	18.00	**1.2**
Cleveland	13.00	16.74	**1.3**
Dallas	10.30	13.50	**1.3**
Miami	13.25	17.40	**1.3**
Indianapolis	11.65	15.60	**1.3**
Phoenix	11.00	15.00	**1.4**
Minneapolis	12.15	17.10	**1.4**
Baltimore	10.75	15.66	**1.5**
Columbus	10.20	15.42	**1.5**
Denver	10.35	16.50	**1.6**
Detroit	12.30	19.80	**1.6**
Seattle	11.70	19.20	**1.6**
San Francisco	12.30	20.70	**1.7**
Chicago	9.50	16.80	**1.8**
Boston	11.10	19.92	**1.8**
Atlanta	10.00	18.00	**1.8**
Houston	9.00	16.50	**1.8**
San Diego	11.35	21.36	**1.9**
Los Angeles	9.40	19.62	**2.1**

Source: http://www.businessinsider.com/uber-vs-taxi-pricing-by-city-2014-10?IR=T accessed on June 12, 2016

Table 2. Comparison between Uber and taxi fares with +20% tip.

More specifically, the most remarkable effect of ride-sharing system has occurred in San Francisco and New York, because Uber and Lyft initially launched in these cities. For that reason, in order to better understand the effect of ride sharing on the traditional taxicab services, we examine the cases of San Francisco and New York here, even though it is still at the early stage for a detailed evaluation. Today, in San Francisco, ride sharing meets an important part of demand for urban travel [2]. In other words, the increase in use of ride sharing led to a 65% decline the number of cab trips in San Francisco in 2 years after Uber's entry into the market, as shown in **Figure 3** [7]. Note that this effect occurred within 2 years.

Similarly, a remarkable impact occurred in New York, even though Uber and Lyft only entered into the market in 2011 and 2014, respectively.[7] **Table 3** reports the change in the market shares of intra-city transportation modes in New York City in 2014 and 2015. Whereas the market share of ride-sharing taxi service was 2% in 2014, it reached 8% in 2015 with a 20,600 ride-sharing taxis.[8] In the same period, the market shares of yellow cabs and other transport services declined 9.52 and 5.76%, respectively, even though the number of medallions remained constant at 13,771. This clearly suggests that taxi users started to substitute ride sharing for cabs. As a matter of fact, the rise in the market share of ride sharing led to an 8% decrease in the number of traditional taxicab trips from 2012 to 2014.[9]

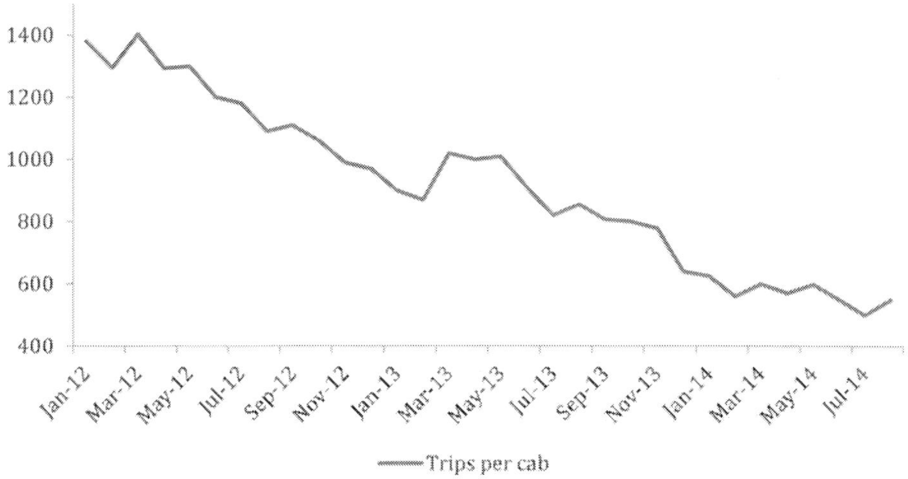

Figure 3. Average monthly number of trips per cab in San Francisco. Source: Taxis and Accessible Services Division report by SF Municipal Transportation Agency, 2014.

	2014	2015	Change
Market share of Uber	2	8	+400%
Market share of green cabs	4	5	+25%
Market share of yellow cabs	42	38	−9.52%
Price of medallions	$1,000,000	$690,000	−31%
Number of yellow taxicabs	13,771	13,771	Constant

Source: http://fivethirtyeight.com/features/uber-is-taking-millions-of-manhattan-rides-away-from-taxis, http://www.economist.com/news/united-states/21661016-does-uber-substitute-cabs-or-attract-new-riders-it-depends-where-you-live-tale?frsc=dg%7Cc

Table 3. Change in the market shares of intra-city transportation modes in New York City.

As different from San Francisco,[10] the presence of ride sharing in New York has led to a radical decrease in the price of medallions as in Chicago [7]. The medallion prices have dropped dramatically, after they peaked at more than $1 million in 2013 but fell to less than $800,000 in 2014 and $690,000 in 2015.[11] Competition from the ride-sharing services such as Uber and Lyft led to a 31% loss in the value of medallions within 2 years. Also, ride sharing has given rise to a drop in taxi profits. Taximeter revenues declined 18.4% from June 2013 to October 2015, as

[7] See for more detail http://fortune.com/2016/03/08/lyft-vs-uber-new-york.
[8] See for more detail http://money.cnn.com/2015/07/21/news/companies/nyc-yellow-taxi-uber.
[9] See for more detail http://www.theatlantic.com/business/archive/2015/07/uber-taxi-drivers-complaints-chicago-newyork/397931.

depicted in **Figure 4** [7]. The findings from these cities are particularly important because ride sharing in urban transport has emerged and developed in those cities. For that reason, the findings from these experiences shed light to the future of relationship between the dynamic ride-sharing and the traditional taxi services. The current developments clearly confirm the rise of ride-sharing system in the big cities of the United States [4].

In other words, the rise of ride sharing improves consumer welfare and thus leads to efficiency in urban transportation as a whole. **Figure 5** depicts the effects of ride sharing on welfare and efficiency from a point of view of microeconomics in a theoretical basis. In the figure, point M refers to the market equilibrium under monopoly market structure, whereas E* represents the equilibrium in a perfect competition market. Note that price is equal to marginal cost (MC) at point E*. In a theoretical sense, under a market equilibrium that refers to monopolistic market structure as at point M, there is dead-weight loss (DWL) represented by the area of MAE*, since the price under the monopolistic market equilibrium is higher than MC. This is the loss in the welfare of consumers who demand the product or service produced by incumbent firm(s) in this market. On the contrary, the point E* refers to an equilibrium in a perfect competition market and there is no loss in consumer welfare at this point. Accordingly, if the market equilibrium shifts from the monopolistic market equilibrium M to the competitive market one E*, consumer welfare increases and efficiency improves, since the users of this good or service pay less for the related good/service and buy more. In such market, resources are efficiently redistributed because there is no DWL. The market is the Pareto-efficient.

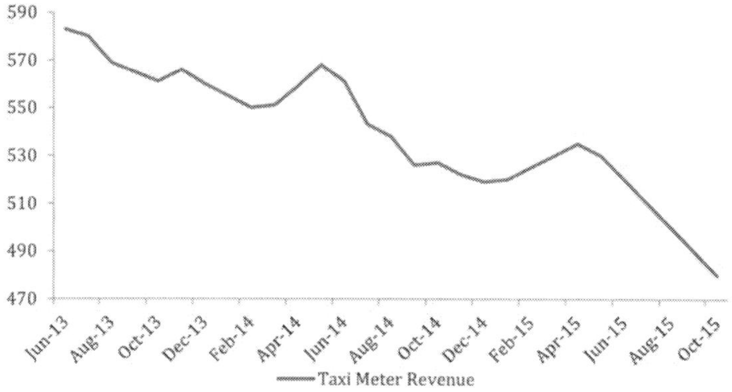

Figure 4. Taximeter revenues from cab trips in New York. Source: TLC. http://www.nyc.gov/html/tlc/html/home/home.shtml.

Of course, it is not easy to find those kinds of markets in the real world. However, this graphical analysis introduces a reasonable theoretical framework for us to better understand the effect of expansion of ride sharing on the taxi services in urban transportation. Accordingly,

[10] Ride sharing has not affected the price of medallions or licenses in San Francisco, because there is no market for medallions as in New York due to the regulatory mechanism in San Francisco.

[11] See for more detail http://skift.com/2015/08/12/new-york-city-taxi-owners-form-group-to-take-on-uber/and http://www.economist.com/news/united-states/21661016-does-uber-substitute-cabs-or-attract-new-riders-it-depends-where-you-live-tale?frsc=dg%7Cc.

we assume that the point *M* represents the monopolistic market structure in the traditional taxicab markets, because the taxicab markets are traditionally subject to strict government regulations such as barriers to entry and this regulatory environment leads to artificial rents or monopolistic profits for medallion owners and a dramatic increase in taxi fares [9, 10]. Because the level of both medallion prices and taxi fares at the point M is higher than the ones in the perfect competition market, it is possible to accept that there is the DWL in the regulated taxicab markets. Accordingly, the area of *MAE* triangle represents the magnitude of loss in consumer welfare in the absence of competitive pressure from ride-sharing service to the traditional taxicab market under regulation. Conversely, the market equilibrium has moved to the point E at the competitive level from the monopolistic market equilibrium through the emergence of ride sharing. As discussed above, both taxi fares and medallion prices have declined after the traditional taxicabs started competing with the dynamic ride-share services. This theoretical analysis, which is consistent with the real market developments, suggests that the rise of ride-sharing services leads to a dramatic decrease in dead-weight loss represented by the area of MAE triangle. We can define this effect as an improvement in economic efficiency in the market for both taxi fares and medallion prices in urban transportation, because medallion prices and taxi fares are close to marginal cost at the equilibrium point E occurring along with the introduction of ride-sharing services into the market. Overall, the results suggest that the rise of ride sharing in urban transportation improves market welfare and economic efficiency.

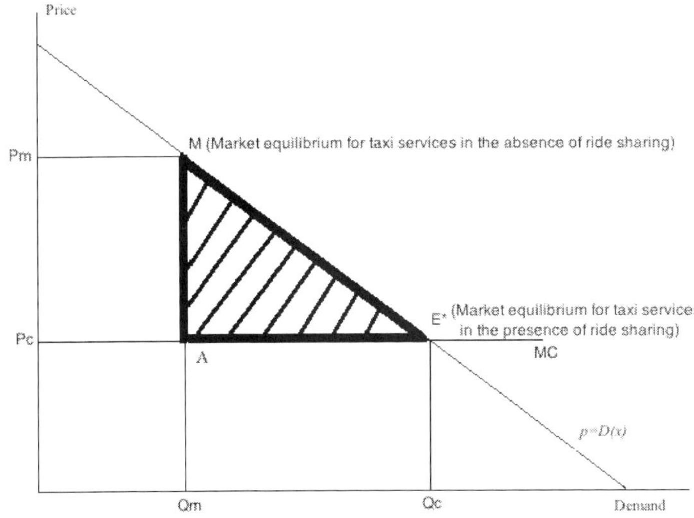

Figure 5. The effect of ride sharing on market welfare and efficiency in taxi services. Source: This figure and analysis is adapted from Ref. [3].

3. Regulation of taxis in urban transport

The conventional academic wisdom is that taxicab market is unique for government regulation [11–14]. Because the main rationale for taxicab regulation is market failure considerations, the aim of regulation of taxis is to ensure economic efficiency by solving market failures. For that reason, taxicab markets have strictly been regulated. Governments have extensively controlled taxi fares and impeded entry to the market. However, this approach is rather controversial today, because, in many cases, regulation of taxis has led to market distortions rather than efficiency. While barriers to entry give rise to monopoly rents for the license owners or interest groups in the market through the regulation-based artificial increases in the value of medallions, price regulation brings about higher fares in the regulated taxicab markets[12] [3, 15, 16].

The failure of regulation and the rise of ride sharing as an alternative intra-city transportation mode have put the inquiry of regulation in taxicab markets into the forefront. While deregulation and competition have been considered as alternative policy measures to regulation [17–19], the ride-sharing system started to threat the justification of regulation. Regarding competition between ride sharing and cabs, even though it is clear that there is an unfair competition that stems from the regulatory asymmetry between taxicabs and ride sharing, the solution of this problem is not to ban or restrict ride sharing through courts or governments, but to regulate ride-sharing services or to deregulate price controls and entry restrictions in the traditional taxicab market. A plausible solution can be to develop a mixed regulatory mechanism that allows competition between ride sharing and cabs in the market. Since it is clear that the presence of ride sharing leads to innovative developments in the market and competitive benefits for users, a mixed strategy including regulation, deregulation, and competition can lead to more efficient market structure than the traditional taxicab markets [4].

For instance, social regulations should continue to be applied for both the traditional taxicab and dynamic ride-share service. In particular, security and safety for users and drivers are rather important. Also, the quality of service and the qualification of drivers should be regulated in both taxi services. Such social regulations cannot be removed from the urban transportation. However, the regulatory experience in the traditional taxicab markets and the results from the rise of ride sharing in urban transportation suggest that economic regulations such as barriers to entry and price controls are not needed anymore in the process of competition between taxicabs and ride sharing. Governments should deregulate economic regulations, because competition between the traditional taxicabs and dynamic ride-sharing services leads to the remarkable improvements in consumer welfare and economic efficiency by decreasing taxi fares and removing artificial or monopolistic rents in medallion prices. On the other hand, social regulations regarding the quality of taxi service in urban transportation including security, safety, and the qualification of driver are still needed.

4. Conclusion

The rise of ride-sharing economy in urban transportation has challenged to the traditional taxicab services. The dynamic ride-sharing startups have introduced alternative taxi services to the users of intra-city public transportation. This radical change among public transport modes

[12] Please see Refs. [9, 10] for more detailed information about the effect of entry and price regulations in the taxicab markets on medallion prices and taxi fares.

within urban transport has led to a dramatic decline in taxi fares and thus an improvement in consumer welfare and economic efficiency in the market as a whole. Rapidly, the ride-sharing services have started to become substitute for the traditional taxicabs. While taxicab services have declined, the ride-share services have raised in urban transportation. On the other hand, some researchers and policy makers started querying the regulation of taxicabs, and some others have argued that the rise of ride sharing negatively affects public transportation.

Developments suggest that the rise of ride sharing will not lead to the death of public transportation. Conversely, ride sharing gives rise to the unique opportunities for the producers and users of taxi services in the intra-city public transportation. However, while competition between ride sharing and traditional taxicabs improves consumer welfare, some new problems occur. But, it is clear that the solution of this problem is not to ban or restrict the dynamic ride-sharing services in urban transportation. Instead, policy makers should prefer a mixed regulatory mechanism. The findings suggest that the best policy option is to deregulate economic regulations such as entry restrictions and fare controls and to continue social regulations including security and safety.

Author details

Tamer Çetin

Address all correspondence to: tcetin@yildiz.edu.tr

Department of Economics, Yildiz Technical University, Turkey

References

[1] Rogers, B. The social costs of Uber. University of Chicago Law Review Dialogue. 2015;82:85–102.

[2] Rayle, L., D. Dai, N. Chan, R. Cervero and S. Shaheen. Just a better taxi? A survey-based comparison of taxis, transit, and ridesourcing services in San Francisco. Transport Policy. 2016;45:168–178. doi:10.1016/j.tranpol.2015.10.004

[3] Çetin, T. and F. Oğuz. The effects of economic regulation in the Istanbul taxicab market. Economic Affairs. 2010;30:59–64. doi:10.1016/j.tra.2011.03.002

[4] Çetin, T. and E. Deakin. Regulation of taxis and the rise of ridesharing, 2016. https://www.academia.edu/28443203/Regulation_of_Taxis_and_the_Rise_of_Ridesharing [Accessed: 2016-09-23]

[5] Shirgaokar, M. and E. Deakin. Study of park-and-ride facilities and their use in the San Francisco Bay Area of California. Transportation Research Record: Journal of the Transportation Research Board. 2005;1927:46–54.

[6] Deakin, E., K. Frick and K Shively. Markets for dynamic ridesharing? Transportation Research Record: Journal of the Transportation Research Board. 2010;2187:131–137.

[7] Wallsten, S. The competitive effects of the sharing economy: How is Uber changing taxis? Technology Policy Institute: Studying the Global Information Economy. Washington, USA, 2015. Available from https://www.ftc.gov/system/files/documents/public_comments/2015/06/01912-96334.pdf [Accessed: 2016-08-23]

[8] Silverstein, S. Uber vs. taxi: these animated charts tell you everything about Uber prices in 21 cities. Business Insider. 2014. Available from http://www.businessinsider.com/uber-vs-taxi-pricing-by-city-2014-10?IR=T [Accessed: 2016-07-25]

[9] Çetin, T. and K. Y. Eryigit. The economic effects of government regulation: evidence from the New York taxicab market. Transport Policy. 2013;25:169–177. doi:10.1016/j.tranpol.2012.11.011

[10] Çetin, T. and K. Y. Eryigit. Estimating the effects of entry regulation in the Istanbul taxicab market. Transportation Research Part A. 2011;45:476–484. doi:10.1016/j.tra.2011.03.002

[11] Guri, D. F. An economic analysis of regulated taxicab markets. Review of Industrial Organization. 2003;23:255–266.

[12] Frankena, M. W. and P. A. Pautler. An economic analysis of taxicab regulation, 1984. Washington, D. C.: Federal Trade Commission. https://www.ftc.gov/reports/economic-analysis-taxicab-regulation [Accessed: 2016-08-23]

[13] Pagano, A. and C. McKnight. Economies of scale in the taxicab industry: some empirical evidence from the United States. Journal of Transportation Economics and Policy. 1983;17:299–313.

[14] Dempsey, P. S. Taxi industry regulation, deregulation, & reregulation: the paradox of market failure. Transportation Law Journal. 1996;24:73–120.

[15] Barlett, A. and Yilmaz, Y. Taxicab medallions-a review of experiences in other cities. Government of the District of Columbia, Briefing Note, May 31, 2011, http://cfo.dc.gov/sites/default/files/dc/sites/ocfo/publication/attachments/ocfo_taxicab_briefing_note.pdf [Accessed: 2015-26-12]

[16] Rufolo, A. M. Low-cost solutions to Portland's traffic problems: congestion pricing and free-market transit. Policy Insight. 1998;105:4–22.

[17] Bekken, J. T. Experiences with (de)regulation in the European taxi industry. In OECD (ed.) (De)regulation of Taxi Industry. Paris: OECD; 2007. p. 31-58.

[18] Hackner, J. and S. Nyberg. Deregulating taxi services: a word of caution. Journal of Transport Economics and Policy. 1995;16:195–207.

[19] Moore, A. T. and T. Balaker. Do economists reach a conclusion on taxi deregulation? Economic Journal Watch, 2006;3:109–132.

Chapter 11

Traffic Congestion Pricing: Methodologies and Equity Implications

Ammar AbuLibdeh

Additional information is available at the end of the chapter

http://dx.doi.org/10.5772/107156

Abstract

Road traffic congestion is recognized as a growing and important urban ill. It occurs in different contexts, takes on many faces and is caused by a variety of processes. It affects both work trips and non-work trips, both passengers and goods flow. It affects the quality of life and the competitiveness of a region. It is an additional cost that arises in the forms of delay, environmental degradation, diminished productivity, standard of living and wasted energy. Congestion pricing can result in winners and losers among different socio-economic groups. However, different studies differ in their conclusions about who wins and who loses because of different assumptions made. This paper reviews the concepts of congestion pricing as a mitigation policy to reduce road congestion and reviews the concept of equity. This paper aims to provide theoretical research that enhances our understanding of congestion pricing policy and the equity implications of this policy.

Keywords: congestion pricing, cordon pricing, equity, cost of congestion, mitigating/managing congestion

1. Introduction

Congestion pricing is an untapped transportation strategy that can reduce traffic congestion, improve air quality and raise the revenue essential to implement needed transportation measures that are effective in improving transportation services and facilities. While experience with congestion pricing is limited, there are sufficient examples and experiences around the world to demonstrate that, when implemented properly, it virtually never fails to be an effective tool to curb congestion. Yet, when initially proposed, it never fails to be controversial. This is in part due to the lack of research on the equity impacts on different socio-economic groups. This is the dichotomy and the dilemma of congestion pricing that every city must face in implementing this new approach to congestion management.

Congestion pricing is the policy of charging drivers a user fee for using certain lanes of roadways that experience congestion, thereby discouraging many drivers from using those lanes and keeping them free of congestion [1, 2]. Congestion pricing captures congestion, operating and capital and environmental costs of vehicle use; therefore, it is considered the best way to deal with congestion and environmental problems [3]. The main purpose of congestion pricing is to mitigate/manage traffic congestion by encouraging drivers to switch to use other modes of transportation, use other routes, or change time of travel (shifting peak-period travel to other off-peak period) [4]. One of the objectives of congestion pricing is to reduce the number of congestion points along roads and hence minimize the length of individual queues that do form. This results in relatively smooth traffic flow with improved fuel economy and reductions in emissions [5].

Several types of congestion pricing have been implemented in several cities around the world. Recent studies in Europe and Asia envision road pricing in the form of area licensing, high-occupancy toll lanes or cordon tolls [6, 7]. This system has been implemented recently in Stockholm. Cordon pricing charges motorists whenever they pass any of the charging points that are located at the entrances of an imaginary zone around a congested area. Charges are flexible, meaning that they vary according to vehicle type, time of day, location and direction travelled [8]. The charges vary between peak and off-peak hours and between weekdays and weekends. This system has proven to be effective in mitigating congestion.

Congestion pricing impacts the travel activities of different socio-economic groups in different ways, albeit in varying ways depending on the circumstances. Cordon pricing has been implemented in some European and Asian cities and has been proposed, but not implemented for North-American cities. All the studies investigated the changes that may occur on people's travel behaviour and hypothesized different ways of redistributing the generated revenues to achieve equity among different travellers based on their socio-economic characteristics. But none of these studies tried to investigate the traveller's preferences in redistributing the generated revenues to achieve equity between different socio-economic groups.

Concerns about equity are raised when considering this system. Travellers who come from outside the cordon have to pay the tolls while residents inside the cordon receive the benefits; also travellers who must travel into and out of the cordon many times during the day have to pay each time. For example, the proposed cordon pricing in Edinburgh, Scotland, was found to be inequitable since people living at equal distances from the proposed cordon were treated differently. Affluent neighbourhoods were exempted from payment as a result of the city's administrative boundaries. On the other hand, it was suggested that less affluent neighbourhoods be subjected to the cordon charges [9, 10]. This example demonstrates the importance of the link between income distribution and spatial equity when designing cordon-pricing systems.

Congestion pricing is a traffic-demand management tool that helps move transportation in the direction of economic and environmental sustainability. At the same time, however, it raises equity issues related to social sustainability as it impacts the travel behaviour of commuters. Equity is operationalized by analysing the progressivity or regressivity of the effects of cordon pricing on groups of travellers based on their socio-economic and demographic factors. Congestion pricing is considered to be regressive or progressive policy if it burdens or favours disadvantaged groups of travellers relative to each other. The interpretation of equity is also based on the broader assessments of transport equity that seek fairness in accessibility and mobility across different socio-economic and demographic groups [11].

In transportation planning, equity is a central element because transportation is perceived as a basic right. That is, access to transportation services is a right to members of all social groups within the society. Thus, many scholars have identified equity concerns as one of the main obstacles to public acceptance of congestion pricing proposals.

2. Approaches for mitigating/managing congestion

Over the years, various approaches have been proposed or implemented to curb traffic congestion and improve the roadway level of service in many countries around the world. These approaches can be considered under either supply management or traffic demand management. Supply management, which is the conventional response to traffic congestion, consists of different techniques such as increasing roadway capacity by expanding or upgrading existing roads or by building new ones. Conventional approaches focus on managing congestion by maximizing the ability of road network to accommodate current and future traffic demand. This approach seeks to maximize the physical usage of road capacity to enhance the levels of service.

A second method is by using different traffic demand management techniques such as encouraging people to use public transit, discouraging peak-period travel, imposing bans on commercial vehicles, parking restrictions and limiting access to congested areas. Another group of traffic-demand management techniques focus on improving the efficiency of the road system to accommodate the same demand at a lower cost. Examples of this approach include imposing charges on road users, high-occupancy vehicle lanes and metering access to highway entrance ramps. **Table 1** describes different approaches to manage/mitigate congestion [12].

The effectiveness of the different approaches in mitigating/managing traffic congestion can be summarized as shown in **Figure 1**. This figure is based on a regional scale and shows that traffic congestion is a consequence of "increased travel demand or inadequate road supply". Traffic congestion mitigation strategies include supply management, demand management and a third alternative which is to do nothing. The "Do nothing" option results in reduced accessibility and mobility and consequently reduces the level of service (LOS). "Increase infrastructure" which is the main action of "supply management" leads to a temporary improvement in the LOS. Put simply, roads are provided, the cost to travel decreases (e.g., higher speeds) inducing more traffic, soon the new road capacity is used during peak-periods, which tend to expand, resulting in traffic congestion and a vicious cycle continues.

For "traffic demand management", different TDM techniques can be implemented to manage/mitigate traffic congestion, including congestion pricing, which is the focus of this research. As a result of implementing "congestion pricing" the LOS will improve. The improvement of LOS also needs to have a two-way relationship with congestion pricing. Pricing is a tool that can be used to maintain an acceptable LOS, requiring "dynamical" adjustments in the toll rates/fees/charges (in real time) to manage the demand and LOS.

Supply management	*Building new infrastructure* This approach aims to increase the roadway capacity. However, it is constrained by a lack of space in dense urban areas as well as funding and environmental restrictions. This approach is expensive to implement and it is considered as the last approach to mitigate traffic congestion. In addition, this approach provides only a temporary solution.
	Modifying existing infrastructure The aim of this approach is to increase the capacity of the roadway by including new lanes, modifying intersections, creating one-way streets and modifying the geometric design of roads. These techniques can benefit public transit as well as car users. However, this approach also requires extensive funding.
Traffic demand management	*Access management* This approach restricts access to specific places or to specific road links. Some of the techniques used in this approach are physical breaks and barriers to block through traffic, permit-based system or traffic bans and ramp metering. This approach is used for safety and is considered most appropriate for reducing the number of cars and increasing the usage of public transit. Some limitations of this approach are that it requires robust enforcement and that road traffic is diverted to other roads creating new congestion.
	Parking management This approach has the potential to modify demand. However, it is under-utilized by many authorities. It can help to reduce demand for automobile travel and, as a result, tackle traffic congestion on the basis of location and time. One limitation of this approach is that the capacity that is freed-up may be filled from through traffic. This approach needs to be supplemented by other approaches to achieve the desired outcomes.
	Improving traffic operations This approach is a cost-effective method to achieve improved travel conditions. The techniques used in this approach include road traffic information system, implementation of dynamic speed, pre-trip guidance and coordinated traffic signal. This approach allows road users to select alternative travel mode or reschedule their trips to off-peak periods.
	Improving public transport This approach is considered a fundamental congestion management strategy. It has the potential to transport more travellers than personal automobiles for a given amount of road space. It can achieve and maintain a high level of access throughout urban areas if the quality of service that it provides is enhanced and sufficient (e.g., safety, comfort, reliability, security) for travellers.
	Mobility management Several mobility strategies can be utilized to mitigate congestion. This approach includes car-pooling, promoting bicycling and walking and large trip generators.

Table 1. Different approaches to manage/mitigate congestion.

The improvements of LOS enhance mobility and accessibility represented as: "transportation" and "land use", respectively. The attributes of mobility/accessibility are applicable to both categories. Transportation and land use need to be coordinated since the trip and location decisions co-determine each other. The spatial distribution of activities co-determines the need for travel and goods movement to overcome the space between the locations of activities. On the other hand, the location decisions of households and firms depend on the

http://dx.doi.org/10.5772/107156

accessibility of locations which results in changes of the land-use system. Under the "transportation" condition, the diagram includes the role of congestion pricing in the decision making process of making a trip by an individual (trip generation/trip distribution, mode choice, traffic assignment/route choice). Congestion pricing has an impact on every step and varies according to the type of congestion pricing scheme, the rates, the area covered and the availability of alternative modes of transport. The distribution of land use determines the location of human activities and consequently the location decisions of investors and users.

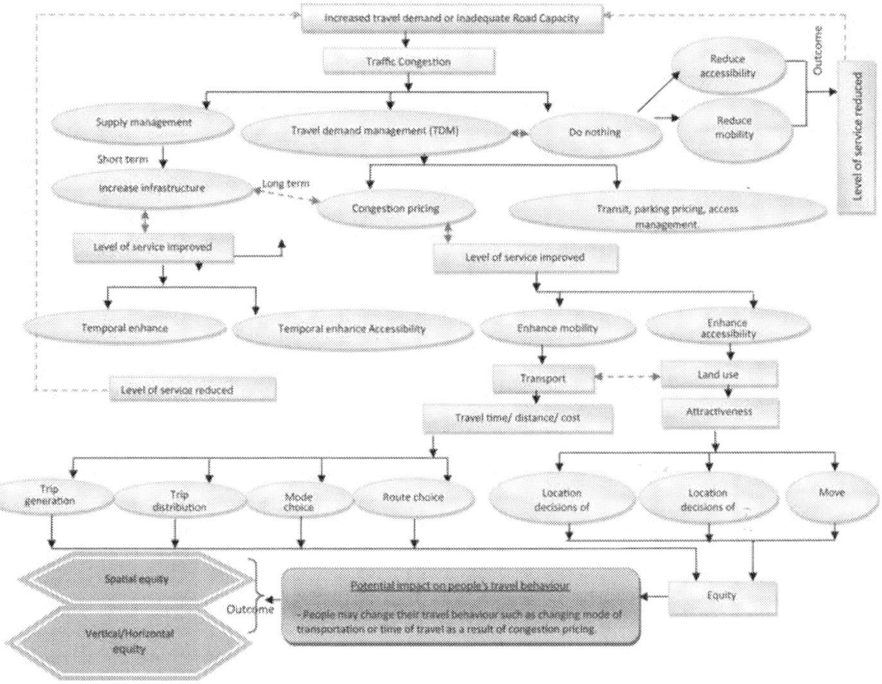

Figure 1. Comparison between different approaches to manage congestion.

To overcome the distance between human activities in space, spatial interactions or trips in the transport system are required. Changes of land use system are associated with the distribution of accessibility in space which co-determines location decisions. In this regard, it is of utmost importance to emphasize the role of congestion pricing in addressing equity concerns (whether spatial or social). Congestion pricing may impact travel behaviour of different socio-economic groups of travellers. People may change their travel behaviour such as changing mode of transportation or time of travel. This questions the impacts of congestion pricing on the equity implications of this policy.

3. Typology of road pricing

Road pricing is a terminology used to include all direct charges imposed on road users including fixed tolls (e.g., toll way) and charges that vary according to the time of the day, location and vehicle size (e.g., congestion pricing) as shown in **Figure 1** [13–16]. Several types of congestion pricing have been implemented in several cities and are identified in the literature. The most implemented forms of congestion pricing projects are shown in **Figure 2** and are presented below:

3.1. Flat-rate toll roads

The aim of imposing fees on travellers in the conventional toll roads is to generate revenues to repay bonds issued to finance the full cost of designing, developing, financing, operating and maintaining the toll way. This system is not considered as a form of congestion pricing since it aims to generate revenue and not to mitigate/manage traffic congestion. Charges are fixed and do not fluctuate according to time or location and can be collected manually or electronically using the transponder technology.

3.2. Cordon pricing

Cordon pricing charge motorists whenever they pass any of the charging points that are located at the entrances of an imaginary zone drawn around a congested area. Charges are flexible, meaning that they vary according to vehicle type, time of day, location and direction travelled [6, 11]. The charges vary between peak and off-peak hours also between weekdays and weekends. Residents inside the cordon pay discounted fees or are exempted from paying the charges. This system is proven to be effective in mitigating congestion.

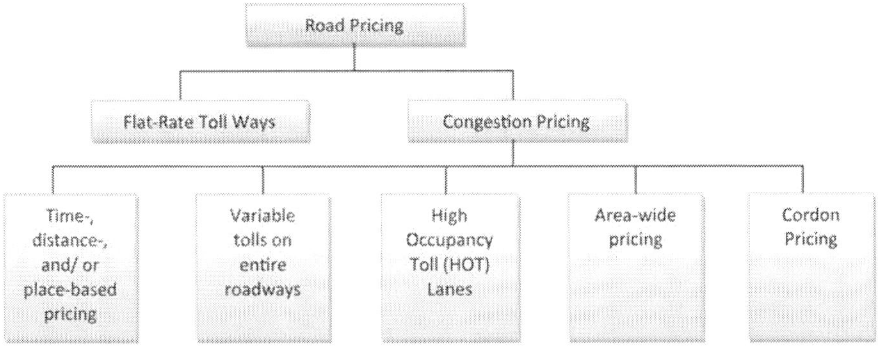

Figure 2. Road pricing typology.

Cordon pricing calls for greater reliance on demand management and on public transportation usage. One aspect of this policy is the restriction of the actual growth of automobile usage through levying a charge on travellers when they cross the priced zone. Cordon pricing in London and Stockholm have been successful in reducing congestion levels and travel time and generating revenues to support transport strategies in these two cities. In addition, the traffic and delay reductions have been maintained over time.

3.3. Area-wide charges

This toll strategy imposes cordon-crossing charges for entering a certain geographic area either by crossing the priced zone or distance travelled (per-km of travel). It is different from cordon pricing in that it charges travellers a fixed fee for traveling across the cordon area for an unlimited number of journeys into and within the priced zone. This system provides a discount for the residents inside the cordon. It is less effective than the cordon pricing at reducing congestion since the fees are fixed and do not change with the number of trips to and from the priced zone; however, it may be perceived to be fairer. Singapore and London implemented this type of congestion pricing in 1975 and 2003, respectively.

3.4. High-occupancy toll (HOT) lanes

This strategy involves variable charges on separated lanes within a highway. It encourages carpooling during peak-periods. HOT lanes are considered a version of high-occupancy vehicle (HOV) lanes. Ride-sharing travellers can use HOT lanes for free or at a discount, while single-occupant vehicles or those that do not meet the minimum passenger occupancy requirements to access the lanes must pay. Typically, transit and emergency vehicles are charged at a reduced rate or are free of charge. All vehicles still have the choice to travel in free, parallel, general-purpose lanes. This strategy is implemented in many cities of the United States. This type of strategy encourages people to use carpooling and the transit system as an alternative to driving alone.

3.4.1. Variable tolls on entire roadways

This strategy depends on changing the charging rate during the peak periods to be higher than off-peak periods. This strategy applies to existing toll roads and bridges to control traffic flow and manage the highway capacity. This aims to encourage drivers to shift to off-peak periods when they use the roads allowing for traffic during the peak periods to flow more freely.

3.4.2. Time-, distance- and/or place-based pricing

This strategy charges travellers based on the distance travelled, location, vehicle type and time of day. The advantage of this system is that it does not require any infrastructure on the ground. It mainly depends on advanced technology where a transponder and a mobile communication device must be installed in each vehicle. This system is implemented in Germany for all heavy-duty trucks operating on the national system. Netherland is in process to develop this system for its entire street and road network.

4. Effect of congestion pricing on travellers based on their socio-economic characteristics

Congestion pricing can result in winners and losers among different socio-economic groups. However, different studies differ in their conclusions about who wins and who loses, as shown in **Table 2**, because of different assumptions made. Earlier studies articulated this issue [17–20] and concluded that low-income or less-flexible travellers (e.g., based on flexibility of working schedule) are considered the worst-off groups and that the approaches of distributing the generated revenues would not make congestion pricing too regressive as the revenue would be used to benefit those who are left worse off. Recent research focused on addressing the importance of equity issues prior to the implementation of congestion pricing scheme and particularly in the design stage [21–23]. Some of these studies proposed a welfare indicator that gives more weight to low-income and disadvantaged groups in terms of cost/benefit ratio. Others proposed a framework for maximizing social welfare (by calculating the optimal road toll) by focusing on spatial equity.

Theoretical studies are different from empirical studies in their conclusions about who wins and who loses. Theoretical studies [24–26] focus on whether congestion pricing benefits low income, high income or both. Different outcomes can be generated based on the assumptions made about different groups in terms of their preferences and travel behaviours. For example, some researchers argue that high-income people believe that their time has a higher value than that of low-income people and hence they (high-income people) benefit the most. In addition, these scholars argue that low-income people live in the suburbs and the work destination for many of them is located inside the city. Therefore, those scholars consider congestion pricing regressive. Other scholars consider congestion pricing as progressive. They argue that the low-income group benefits the most from congestion charging since they more often use public transport; hence investing the generated revenue in improving this mode of transportation benefits this group. On the other hand, quantitative (empirical studies) studies have been conducted to study congestion pricing equity for some cities [27–30]. Most of these studies conclude that high-income people are more negatively affected by a congestion pricing system since they drive more than low-income people. Also, high-income people tend to live in areas with poor access to public transportation and hence will be more affected by this policy.

Differences in conclusions about who may win and who may lose can be attributed to two main reasons, which are researcher's background and the methods used in the analysis. Geographers have long held interest in addressing the challenges that urban commuting poses to society [41, 42]. Geographers focus on spatial dimensions when addressing commuting problems. The spatial separation of people's origin (e.g., home) and destination (e.g., work) and the prevailing urban structure influence people's commute. Several geographers, as a result, focused on addressing, theoretically and empirically, the connection between travel patterns and land use [43]. Therefore, in assessing the equity implications of transportation in general and congestion pricing in particular, geographers, as well as transport planners, look at those who may be disadvantaged (e.g., because of age, gender, disability, income) with respect to transportation. In terms of congestion pricing, it is important for geographers to know where people live since some neighbourhoods may

http://dx.doi.org/10.5772/107156

burden charges more than others. Traffic engineers, on the other hand, are concerned more with system efficiency more than system equity [44]. They seek to enhance transportation infrastructure to increase roadway capacity and to improve traffic flow to maximize tangible benefits for a given cost [44]. Economists tend to group people based on their income level and are concerned with the distribution of costs and benefits among these groups to assess the equity implications of transportation and congestion pricing.

Reference	The overall effects	Winners	Losers
[1]		People with full-time employment, those from higher income neighbourhoods; and this holds true when the population is broken out by gender, age group, household size and occupational class	Professionals, those who live in one- and two-person households and those who are aged 65 and older would be disproportionately affected; those who work in manufacturing would be less affected
[31]	A proper allocation of revenue such as investing in public transportation network and infrastructure or reduce regressive local tax would have better effect on low-income drivers when implementing charging scheme		Low-income travellers are harmed by the imposition of the charging scheme hence the time saving they would gain would not compensate what they pay
[26, 32]		High-income travellers	
[33]		Car travellers changing their mode of transportation to public transit if time savings of these facilities are substantial	
[34]	Road pricing would not be too regressive as the revenue would be used to benefit those who are left worse off		Those who work in the charged area, drivers with low values of time, solo drivers or travellers in vehicles with lower occupancy, travellers that do not have time flexibility and those who cannot switch to other modes of transportation to avoid charges
[24, 27, 31, 35]	Congestion pricing will be regressive as the monetary value of time for high -income travellers is greater than those of low-income and hence they are more willing to pay the charges as they feel that their time gain is worth the fees		High-income travellers working in small economic margins suffer more from congestion pricing as they cannot avoid the charges levied during peak hours as they have inferior possibilities to decide their time of work

Reference	The overall effects	Winners	Losers
[36]	Identified few regressive effects of cordon pricing in London since high-income travellers use their own cars in their commuting more often than low-income travellers		Even though low-income travellers benefit as a group from this policy, yet low-income individuals who cannot switch to other modes of transportation and still needed to use their cars would be severely affected
[37]		Travellers valuing the time savings higher than the fee Persons now finding it "profitable" to undertake a trip (or change trip timing, route or mode choice), even with a fee, because the travel time will be reduced Public transport passengers experiencing time savings Commercial enterprises which undertake substantial transport activities	travellers valuing the time savings below the fee, but having only unattractive travel alternatives Persons abstaining from travel or changing to less attractive travel times, routes or modes to avoid fee Persons experiencing congestion on a road or on public transport, caused by persons who have changed travel behaviour to avoid fee
[38]	Road pricing is progressive rather than regressive as low-income group benefit more as they tend to use public transportation more often The final effect of this system would be progressive if the generated revenues are distributed on improving public transportation, enhancing cycling and walking and enhancing traffic calming	Those who currently use other modes of transportation than cars for their daily commuting Those who have a high value of time	Those who encounter increase in travel cost or take more time as a result of using alternative modes of transportation as well as those who have lower value of time and continue to travel by cars and hence time benefits are not offset by the cost of the charges
[39]		Public transportation users would all be winners because they reap the benefits of low road congestion and the improvements in public transportation network without paying the charges	Travellers that pay the standard charges will be the losers because they will most likely experience reduced road congestion and increase in travel speed that are not sufficient to offset the financial loss of the fees Those who transfer to use public transportation as they are not traveling by their preferred mode of transportation

Reference	The overall effects	Winners	Losers
[23]	Progressivity or regressivity of such a policy is mainly related to the choice of the method of allocating the generated revenue Neglecting the refund scheme, the welfare effects of the policy are borne largely by high-income travellers as they are predominantly car users and therefore the scheme itself tends to be progressive		Low-income travellers who use their cars also bear a high burden
[21]			High-income travellers are more likely to live in the suburban areas outside the city core in areas where public transport is poor
[19, 21, 28–30, 40]			High-income people are more likely to drive more than low income people and tend to live in areas with poor access to public transportation therefore they will be more likely affected by congestion pricing policy

Table 2. Winners and losers when road pricing is implemented on an existing road system.

The second reason is the methods used in the analysis. Different empirical approaches and analytical techniques were used in addressing equity in terms of congestion pricing. These techniques can be grouped into three different categories which are mathematical models, GIS and key-interviews and surveys. Mathematical models are built to address different aspects of equity. Numerous data are used such as origin/destination, travel time, gender, income, location, car ownership and family situation and occupational status. However, the results of some studies that used this approach are contradictory. The second approach is the use of geographic information system (GIS). This approach was used by many geographical scholars to address the impact of transportation on the environment and on the society as a whole. GIS is used in commute studies because it has the capability of handling the spatial data that is important in road network modelling process which is vital for computing streets-based measures of both distance between zones and travel time [45]. Studies that used GIS to address transportation and congestion pricing equity used several types of data such as place of residence, place of work, mode of choice, socioeconomic characteristics and commuting flows. The third approach used is a combination of key interviews, surveys and focus groups. This approach is an excellent method to collect information about opinions, meanings and experiences. It is used frequently as a flexible tool to obtain in-depth information from the respondents. However, this approach is restricted by possible bias introduced by the presence of the researcher and researcher's data interpretation and respondents' personal differences in articulation [46].

5. Overview of the concept of equity

The determination of just distribution of rewards, resources, rights, duties, obligations and liabilities or costs; and the allocations of positive and negative outcomes within social systems are of considerable interest to social scientists. Equity is the value of being equal or fair. As equity is concerned with the fair distribution of society's resources among individuals and groups, it is extensively received as positive and as an objective in social policy. Moreover, it has become a significant criterion in assessing public policy and programs dealing with the optimal use and distribution of resources [15, 47]. Many social policy definitions include aspects of equity, equality, justice or fairness (see the definitions given in Refs. [15, 47]). Equity is frequently identified as "distributional fairness"; as its main concern is "who gets what" and with "who pays" [48, p. 19]. "Equity objectives can be identified in four main sets: guaranteeing minimum standards; supporting living standards; reducing inequality; and promoting social integration" [47, p. 48].

To achieve equity, the distribution of costs and benefits, whether monetary or non-monetary, must be seen by society to be fair and just depending on an array of criteria. Thus, a policy can be described as equitable if it satisfies a normative standard of fairness [15]. However, reaching an agreement on what constitutes equity is almost always context-specific. Therefore, as Murray and Davis [49] argue, the definition of equity requires a set of universally accepted norms; while its practice and interpretation are both comparative and specific.

5.1. Theories and principles of equity

The "egalitarian principle" is the starting point of social justice theory that calls for equality among individuals in a society and equality is understood as the treatment of people as equals. However, applying the egalitarian principle is difficult. For example, a society may try to achieve an egalitarian distribution of wealth by ensuring that equal inputs (food, education, ...etc.) are offered to each individual. However, this fails to take into consideration the difference among different members of the society in labour as some individuals may convert inputs into greater wealth generation than others. In reality, what comes into view as an egalitarian distribution of wealth may at the end lead to inequality. On the other hand, in *Distributive Justice*, Rescher [50] argues that society should commit unequal inputs to accomplish equal rights for members. He defines rights as the traditional personal freedoms and equal opportunity to education and employment. His solution starts by assuring all members a minimum equal standard of living that he referred to as "utility floor" that points to the minimally acceptable share of necessary goods, such as food and shelter. Beyond this point, he believes that in order to motivate individuals to boost production and consequently, to stimulate the furthermost good for most of the members in a social system, output inequality in terms of inequality of wealth and circumstances should be allowed in society. Without this inequality, which he describes as incentive, scarcity may take place and may hinder the achievement of the "utility floor" for all individuals.

Although the egalitarian principle suggests that resources should be distributed equally among citizens without any segmentation, Osterle [47] argues that egalitarian principles might be regarded as appropriate in some areas of social policy, while in others they may be regarded as inappropriate. For example, these principles are appropriate in social policy regarding child benefits or education aiming at equal opportunities, while these principles do not seem appropriate when distributing equal shares of care without taking into consideration different levels of disability.

Despite the continuing debate, a revolution in our collective understanding of the concept of equity has taken place as many authors have adjusted their earlier definitions taking into consideration the differences in needs and abilities of members of society. Equity theorists are occupied with determining the principles of distributive justice under different social settings and with identifying when such principles are perceived as fair or just by individuals within the social system. Focusing on outcomes or procedures, equity theories imply principles of how equity should be defined and suggest principles to be applied in different contexts. On the other hand, empirical equity studies emphasize equity viewpoints and equity judgments or on testing certain equity interpretations. These are often derived from theories of justice or equity judgments. While, evaluating particular interpretations of equity has received significant attention by scholars, fewer studies consider how concerns about equity are translated into social policy practice. Although, there is increasing information about the distribution of costs and benefits according to particular interpretations of equity, a lack of evaluation research is noticed dealing with "whether and to what extent these interpretations reflect explicit or implicit social policy objectives, or whether there might be competing equity concerns" [47, p. 49]. As Osterle [47, p. 56] further notes, "no attempts have been taking place to study the complete range of such questions and to propose a conceptual and theoretical framework to illumine how institutions distribute costs and benefits". This has led to a significant gap between "searching for ideal concepts of equity and investigating societal outcomes".

On the other hand, equity concerns in social policy are often determined by three dimensions: what is to be shared (resources and burdens); among whom (the receivers); and how (the principles). Taking into account these three dimensions is a means for the illumination of equity objectives that are in many cases vague or not well-defined [47]. Campbell [51, p. 3] wrote three decades ago: "The question of how to make operational the equity principle will become an increasing concern. At the heart of these concerns will be defining equity, developing measures of it, collecting and interpreting relevant data and developing policies responsive to it". Campbell's questions and concerns still occupy many researchers from different disciplines particularly, human geographers, planners and economists.

With regard to evaluating equity in social policy, it is important to emphasize that a lack of specifically and clearly defined equity objectives is a key difficulty when assessing equity concerns. However, three different sets of approaches can be distinguished in the literature. First, theories of justice are considered as the point of departure to evaluate equity in social policy. However, the issue of equity is at the core of the debate about these theories. Some scholars emphasize issues of social policy as healthcare, for example, by searching for the content of a just distribution of resources [47]. Le Grand (1991 qtd in Osterle, 2002) [47, p. 49] evaluates equity by looking at the range of opportunities and choices that exists for individuals in a society. He states that "*situations where one person is disadvantaged relative to another due to factors beyond either's control are commonly judged inequitable; situations where the disadvantage arises because of differences in individual choices freely made are not*". Within the same context, Daniels [52, p. 57] states that "shares of the normal range will be fair when positive steps have been taken to make sure that individuals maintain normal functioning, where possible and that there are no other discriminatory impediments to their choice of life plans". Although, theories of justice are considered by scholars following such approaches as the point of departure in evaluating equity in social policy, the prospective of these approaches in empirical work remains limited. This is due to constraints in translating ideas of welfare economics, for example, to assessment applications. The second set of approaches emphasizes equity beliefs, expectations and judgments. Furthermore, causes and effects of such judgments are also emphasized. This approach is useful in the descriptive examination of equity and is considered as the foundation for explanatory studies regarding judgments and beliefs by individuals. However, the main critique within the debate about such an approach is the taking apart from normative, philosophical ideas of justice. The third set of approaches emphasizes the analysis of outcomes. It highlights the extent to which empirical distributions respond to definite interpretations of equity. In healthcare, there are several studies that address such questions. For example, some scholars examine equality in the distribution of health, while others examine the distribution of public expenditure and outcome for a variety of policy areas such as health and social services. In many cases, the analysis is based on five different interpretations of equality: equal public expenditure, equal final income, equal use, equal cost and equal outcome. Equity studies are rather rare in other areas of social policy; a number of studies in long-term care are exceptions [47].

5.2. Equity implications of congestion pricing

In transportation planning, equity is a central element because transportation is perceived as a basic right. That is, access to transportation services is a right to members of all social groups within the society. Thus, many scholars have identified equity concerns as one of the main obstacles to public acceptance of congestion pricing proposals. Indeed, a claim that potential equity impacts have not been carefully examined makes the implementation of congestion pricing very slow.

Equity is a major concern that is raised prior to and after congestion pricing implementation. This is due to imposing charges on access to roadways that were previously free, which may be perceived to harm especially lower income groups because they will either have to pay the fees or be priced off the roads. Advocates of congestion pricing argue that implementing this system is more equitable and less regressive than the current systems (e.g., motor fuel taxes, property taxes, license fees and registration fees) to manage the use of roads as well as to fund transportation improvements. In short, drivers who contribute most to road congestion under a congestion pricing scheme will pay more for using transportation facilities. Critiques of the current financing system in North America suggest that it is regressive and not equitable since low-income drivers pay a higher proportion of their income for transportation fees and taxes than the high-income drivers. In terms of congestion pricing, some critics argue that congestion pricing is unfair, particularly to lower income people who need to drive, because it imposes "double charging", given that drivers already pay registration and

http://dx.doi.org/10.5772/107156

fuel taxes. Moreover, some drivers pay more than others which raises debate about what pricing is equitable and how modifications can be fair and advantageous to the drivers. Another dimension of equity of congestion pricing is its ability to reduce air pollution. This is particularly beneficial to low-income neighbourhoods that are sometimes located in the vicinity of major roads and other transportation facilities.

Within the economic literature of equity and congestion pricing, the work of Rawls [53] noticeably renewed the approach of justice within the analysis of transport policy. According to the theory of Rawls leads to the identification of three dimensions of equity directly relevant to the transport realm and its pricing. These are shown in **Table 3**. Within these dimensions, there are four main points that should be highlighted: First, horizontal equity implies that members of the same group or same circumstances should be treated identically. Horizontal equity is concerned with allocating public resources equally among like individuals and like classes; in other words, it is concerned with fairness between persons and groups with equal resources, abilities and needs. According to this definition, equal persons or groups should get what they pay for and pay for what they get. They should be treated equally, tolerate equal cost and receive the same shares of resources.

Second, vertical equity is concerned with the distribution of differential effects on individuals or groups that differ by socio-economic factors such as income; in other words, it is concerned with the treatment of persons and groups that are dissimilar. Based on that, the allocation of costs and benefits should reflect individuals' needs and abilities.

The third and fourth principles deal with motorists as actors. More specifically, the third principle is that those who contribute to a social cost should pay for doing so; this is referred to in the literature as the "cost principle". Fourth, those who receive social benefits pay for them; this is referred to in the literature as the "benefit principle".

In terms of the use of any potential profits from road pricing schemes, there is a difference of interpretation between horizontal and vertical equity. Horizontal equity implies that profits should be devoted to roadway projects or rebated to vehicle users as a class, but this condition is reduced or removed if the analysis distinguishes the need for users to recompense for the external costs they entail. In contrast, vertical equity justifies employing revenues to the advantage of under-privileged people, such as low-income drivers as a class and non-drivers. Litman [54] notes that this can be accomplished by utilizing resources to benefit lower income drivers or to develop transportation alternatives such as transit, bicycling and walking; and to furnish public services that benefit low-income earners in the society.

Equity could be in terms of who pays/who benefits (car users, transit, non-motorized), income equity (need to look for poverty levels and whether charging is more regressive than other taxes), gender equity (male/female), geographic equity (urban/rural/suburban), its relationship to other charges and fees (property taxes, how transportation projects are funded), accessibility to travel alternatives (if I leave my car to avoid charges, are there reliable transit alternatives), business equity (impact on businesses in areas with congestion charges versus those that are not impacted). Equity reflects the changes in the allocation of impacts (costs and benefits) across socio-economic groups, resulting from the introduction of pricing decisions, relative to the existing allocations.

Dimensions of equity	
Spatial equity	• "Corresponding to the 'principle of liberty', in which the society must guarantee everywhere the access rights to the goods and the services" [55]. • Benefits of transport strategies and services should be equally distributed particularly on those with special needs; lower income residents, elderly and disabled people, those who do not own cars and those living in underprivileged areas. • This dimension of equity is concerned with avoiding worsening accessibility, the environment or safety for any of the social groups. • Social inclusion is a related issue concerned primarily with accessibility (or lack of it) for those without a car or whose mobility is prejudiced.
Horizontal equity	• "Corresponding to the 'principle of equal opportunity', which concerns the equal treatment between users and the user-pays principle" (PATS, 2000, p. 59). • Horizontal equity implies that all people in a given group are equal and should enjoy equal social, political and economic rights and opportunities. It simply means similar distribution of costs and benefits to individuals within a group. • A transport policy is horizontally equitable if similar individuals are provided with equal opportunities or are made equally well off under the policy. • Horizontal equity assumes that "like should be treated alike." It is often interpreted to mean that individuals should "get what they pay for and pay for what they get". • Road pricing revenues should be dedicated to road improvements or to provide other benefits to people who pay the fee. • Horizontal equity implies transferring benefits from one group (those who pay the fee) to another (those who do not).

http://dx.doi.org/10.5772/107156

Dimensions of equity	
Vertical equity	• "Corresponding to the 'principle of difference', which explicitly takes into account the inequalities and its consequences as regards transport." (PATS, 2000, p. 59). • Vertical equity is concerned with the treatment of individuals and classes who are unalike. Therefore, the distribution of costs and benefits should reflect people's needs and abilities. • It often differentiates between groups based on ability to pay, which is typically measured by an individual's income or wealth. • A transport policy is progressive or regressive depending on whether it favours or burdens, based on some measurable criteria and disadvantaged individuals relative to others. • While these costs and benefits are often expressed in monetary terms, they could be measured in other ways as well. • Vertical equity often requires that disadvantaged people receive more public resources (per capita or unit of service) to accommodate their greater need than those who are advantaged. • It justifies employing revenues to the advantage of underprivileged people, such as low-income drivers as a class and non-drivers. Litman (2007) notes that this can be accomplished by utilizing resources to benefit lower-income drivers or to develop transportation alternatives such as transit, bicycling and walking; and to furnish public services that benefit low-income earners in the society.

Table 3. The three dimensions of equity based on Ref. [53] Rawls' theory.

This leads us to the problem of deciding how to make comparisons among different social groups within the society. The economics literature classifies members of the society based on their income or their place of residence or work, while the planning literature consider those who may be disadvantaged with respect to transportation because of disability, age or gender.

However, congestion pricing must also consider where people live, as some neighbourhoods may experience greater burden than others because of the way in which we implement congestion pricing.

In conclusion to the above discussion, one may argue that there is no easy answer available to the question that is often raised, "Is congestion pricing equitable?" There is not a theory of equity but multiple meanings of the concept proposed by human and social sciences. And the answer to this question largely depends on how we measure equity and how we define groups, the details of the site and lastly, to what we judge against congestion pricing. However, in an attempt to answer the above question, the literature about congestion pricing and equity has been reviewed and one can suggest the following conclusions regarding this issue:

First, an equity evaluation must carefully consider socio-economics, demographics as well as location. The distribution of residents, job opportunities and other vital destination has, to a great extent, a significant impact on equity implications for all types of congestion pricing. Cordon pricing, for example, may be progressive, regressive or neutral based on the place of residence of low-income people.

Second, an important factor for the net impact of congestion pricing is how revenues are used. Differences in this respect reduce differences in other factors such as values of time. Having to pay for what was freely available and the risk of exclusion for low-income social groups for the extra cost of driving causes political hostility. Thus, from an economic perspective, spending revenues in ways that benefit low-income and other transportation-disadvantaged social groups will make congestion pricing more likely progressive rather than regressive. This is largely dependent on how congestion pricing is implemented. However, if revenues and benefits are distributed equally within society, congestion pricing may be taken as a whole as regressive. On the other hand, even with spending revenues in ways to benefit low income, it is still possible that some members will still be disproportionately burdened.

In terms of equity impacts, the literature on road pricing has focused mainly on income equity issues and to a lesser extent on spatial equity. In general, the three congestion pricing projects that were implemented in the Asian city (Singapore) and the two European cities (London and Stockholm) gave equity only limited attention and evaluation. When charges are imposed on travellers, these result in perceived road user's "winners" and "losers". This is attributed to the way that travellers value time savings, where some road users value these savings more than the fees they pay. The losers, who are tolled off, may change their travel routes, shift to off-peak times, change the mode of transportation, shift to carpool or make fewer trips. In Singapore, gainers from congestion pricing project were found to outnumber losers 52 to 48% [50]. Also, after implementing congestion pricing in Singapore it was found that residents outside the priced zone considered this project as negative while residents inside the priced zone considered it positive. The enhancement of public transit before implementing congestion pricing can be considered a way to achieve equity between different income groups. In Stockholm, transit service was extended by 7% by adding 16 new bus lines, additional departure for train lines and new park-ride facilities 4 months before the start of the tolling.

http://dx.doi.org/10.5772/107156

Two commonly suggested ways to mitigate the risk of negative impacts of congestion pricing on low-income and disadvantage groups are found in the literature. The first approach is to distribute the revenue generated from congestion pricing through public works and in particular, on the public transit system to create better options not to drive and to ensure that project benefits flow to those most disadvantaged individuals by congestion pricing. Other ways identified in the literature on redistributing the generated revenue are through tax credits and credit-based systems to ensure that redistribution is made on an individual basis. However, none of these ways were tested or implemented in reality; therefore, their effectiveness is difficult to judge. The second approach is discounts and exemptions for disadvantaged (e.g., disabled persons) and low-income individuals, vehicles or types of trips. This approach leads to a less expensive congestion pricing system. However, the incentives to discourage drivers to travel on congested roads will be reduced if a large number of people get discounts or exemptions.

The last point on promoting equitable outcomes is that a region seeking to implement congestion pricing should look at measuring and assessing equity in the early phases of the planning process. Most importantly, a proposal of congestion pricing should be tested through modelling to determine who are more likely to pay the charges and whether the situation of the low income and transportation-disadvantaged social groups will be worse off with the proposed project. Furthermore, public participation should be facilitated so members of the society affected by this project are aware of it and also are given the chance to offer suggestions. Lastly, even after the implementation of the congestion pricing, equity has to be monitored and changes should be made every so often to the system if the early tools to endorse equitable outcomes are not achieving their goals. It would be also functional to develop an "equity audit tool" to facilitate this process.

In conclusion to the above discussion, the concept of equity is subject to broad interpretation. This notion deals with principles that identify the fair or just distribution of resources among members of the society. Because the formation of these principles entails ethical and subjective judgment, the study of equity is burdened with definitional mystification and "pluralism". In general, equity definitions stress the significance of a fair distribution of benefits and burdens. Furthermore, accurate definitions of equity are rare in both policy making and policy evaluation. Thus, reaching an agreement on what constitutes equity and the fairness of a specific distribution is almost unattainable which makes the concept of equity a complex one. The difficulty in defining *equity* as descriptive and normative has made the theoretical literature on equity very debatable.

Lastly, it may be useful here to clarify the link between theories and principles of equity ("economic theory", "social justice theory", "Rawls' theory") that are introduced in this chapter. A theory of social equity was developed and positioned as the "third pillar"; in addition to economy and efficiency for transportation planning including road pricing. In terms of the theoretical progress of equity in the last two decades, the work of Rawls provided a language and a road map for transportation planners to understand the complexity of the subject and to integrate notions of fairness, justice and equality in their planning.

Rawls derives his two principles of justice: "the liberty principle" and "the difference principle" from his theory that is known as "Justice as Fairness". He claims that adopting two such principles organizes the distribution of economic and social benefits across society. The difference principle justifies unequal distribution of goods only if those inequalities are to the advantage of the worst-off members of society. With the emergent focus on congestion pricing, concern is rising about whether congestion-based charging policies can be designed in an equitable way. Therefore, Rawls' theory, particularly the difference principle, can help planners to develop criteria for assessing public policy and programs dealing with the optimal use and most importantly, the distribution of resources. The next chapter has further discussion about theories and principles of equity with regard to congestion pricing.

6. Conclusion

Congestion continues to increase in many cities around the globe and the traditional approaches of expanding transportation infrastructure or building more roads to operate at minimum congestion at all times will not be a solution due to financial and environmental reasons. Congestion pricing has become an increasingly practical option implemented in various forms for managing congestion, protecting the environment and raising revenue for investments in transportation. It has been easily implemented in recent years because of the advances in technologies that make it achievable to charge motorists as they drive. Although, transportation planners and policy makers are considering congestion pricing as a promising alternative to mitigate/manage congestion, it has thoroughly faced an unreceptive public and political environment. While few cities succeeded in implementing different schemes of congestion pricing, yet many proposals were discarded based on equitability concerns. In general, equity has been given limited attention and evaluation when cordon pricing was implemented in different cities around the world.

However, the equity of pricing schemes is a major concern among the public and elected officials prior to and after congestion pricing implementation. This is due to charges being imposed on access to roadways that were previously free, a change that may harm different socio-economic groups such as low-income travellers, because they will either have to pay the fees or be priced off the roads. The issue of equity is at the core of the debate in social science, particularly with regard to assessing equity in social policy. Several reasonable and conflicting notions of equity exist and, as identified in, this is related to the fact that there are several impacts to be considered. But, at the same time, many of these are difficult to measure and there are numerous ways to classify "winners and losers". There is not an accepted and commonly used manual for evaluating equity in transportation policies.

In conclusion to the above discussion, one may argue that there is no easy answer available to the question, "Is implementing congestion pricing policy is equitable?" There is not a theory of equity but multiple meanings of the concept proposed by human and social sciences. However, the answer to this question largely depends on how we measure equity and how we define groups. Taking into consideration all aspects of equity is impossible. However, in an attempt to answer the above question, one can suggest the following conclusions regarding this issue:

http://dx.doi.org/10.5772/107156

It is concluded that the most important factor for the net impact of congestion pricing is how revenues are used. Differences in this respect reduce differences in other factors such as values of time. Having to pay for what was freely available and the risk of exclusion for impacted socio-economic groups for the extra cost of driving causes political hostility. Thus, spending revenues in ways that benefit people from low-income neighbourhoods and other transportation-disadvantaged social groups will make congestion pricing more likely progressive rather than regressive. On the other hand, even with spending revenues in ways to benefit these groups, it is likely that some members will still experience a burden.

Although utilizing revenues to improve transit services is considered to be an effective strategy for increasing equitable outcomes, still not all transit is created equal and it is not considered by many as a viable strategy for addressing equity concerns. Investments in various modes of transportation and different neighbourhoods may have different impacts.

Different approaches can generally evaluate the measurement of equity in transportation in many ways. The difficulties of these evaluations are greater when applied to congestion pricing than other forms of transportation demand management or financing schemes such as taxation. This is due to the fact that the range of congestion pricing impacts is quite larger. The evaluation of equity for congestion pricing policy can be complicated due to the many variables involved. For example, the incidence of congestion pricing relies on location. Therefore, the places where individuals in the same income groups live, worship, work and shop are a critical element of how these individuals experience congestion pricing.

Other essential factors are cost, convenience, presence and cost of alternatives to driving. Equity implications are different if individuals can switch from using their own automobiles during congested time to other modes of transportation such as public transit, walking or cycling. In addition, comparing equity implications of cordon pricing in different cities around the world is fundamentally impossible because of the many other factors that may play a significant role in the outcomes.

Author details

Ammar AbuLibdeh

Address all correspondence to: a.abulebdeh@uaeu.ac.ae

Department of Geography and Urban Planning, Facility of Humanities and Social Sciences, United Arab Emirates University, Al Ain, United Arab Emirates

References

[1] Abulibdeh, A., Andrey, J. and Melnik, M. (2015), Insights into the Fairness of Cordon Pricing Based on Origin–Destination Data. Journal of Transport Geography. Vol. 49, pp. 61–67.

[2] Black, R. W. (2010). Sustainable Transportation: Problems and Solutions. Guilford Publication, New York.

[3] Kitchen, H. (2008), "Financing Public Transit and Transportation in the Greater Toronto Area and Hamilton: Future Initiatives". Completed for: Residential and Civil Construction Alliance of Ontario, 25 North Rivermede Road, Unit 13, Vaughan, Ontario.

[4] FHAW (2006), "Issues and Options for Increasing the Use of Tolling and Pricing to Finance Transportation Improvements", Final Report, Work Order 05-002.

[5] Abulibdeh, A. (2012), "Equity Implications of Cordon Pricing in Downtown Toronto", A Thesis Presented to the University of Waterloo in Fulfillment of the Thesis Requirement for the Degree of Doctor of Philosophy in Geography. Waterloo, Ontario, Canada.

[6] Hyman, G. and Mayhew, L. (2002), "Optimizing the Benefits of Urban Road User Charging." Transport Policy, Vol. 9, No. 3, pp. 189–207.

[7] de Palma, A., Kilani, M. and Lindsey, R. (2003), "Congestion Pricing in the City: Metropolis Simulation Results", Presented at 50th Annual Meeting of the Regional Science Association International, Philadelphia, PA.

[8] NCHRP (1994), "Road Pricing for Congestion Management: A Survey of International Practice", A Synthesis of Highway Practice, Synthesis 210, Transportation Research Board.

[9] Raje, F., Grieco, M. and McQuaid, W. R. (2004), "Edinburgh, Road Pricing and the Boundary Problem: Issues of Equity and Efficiency", Stirling and Glasgow: Scottish Economic Policy Network.

[10] Rouwendal, J. and Verhoef, T. E. (2006), "Basic Economic Principles of Road Pricing: From Theory to Applications", Transport Policy, Vol. 13, pp. 106–114.

[11] Levknson, D. (2010), Equity Effects of Road Pricing: A Review. Transport Reviews, Vol. 30, No. 1, pp. 33–57.

[12] Black, R. W. (2003), Transportation: A Geographical Analysis, The Guilford Press, New York.

[13] Evans, E. J., Bhatt, U. K. and Turnbull, F. K. (2003), "Road Value Pricing Traveler Response to Transportation System Changes", Chapter 14: Road Value Pricing. TCRP Report 95, Transit Cooperative Research Program. Transportation Research Board. Washington, D.C.

[14] Ungemah, D. (2007), "This Land is your Land, This Land is my Land: Addressing Equity and Fairness in Tolling and Pricing". Transportation Research Record: Journal of the Transportation Research Board, No. 2013, Transportation Research Board of the National Academies, Washington, D.C., pp. 13–20.

[15] Ecola, L. and Light, T. (2009), "Equity and Congestion Pricing: A Review of the Evidence", Technical report, Published 2009 by the RAND Corporation, Transportation, Space and Technology, Santa Monica, CA.

[16] FHWA (2006), "Congestion Pricing: A Primer", Publication Number: FHWA-HOP-07-074.

[17] Richardson, H. W. (1974), "A Note on the Distributional Effects of Road Pricing", Journal of Transport Economics and Policy, Vol. 8, pp. 82–85.

[18] Cohen, Y. (1987), "Commuter Welfare under Peak Period Congestion Tolls: Who Gains and Who Loses?" International Journal of Transport Economics, Vol. 14, No. 3, pp. 239–266.

[19] Fridstrom, L., Minken, H., Moilanen, P., Shepherd, S. and Vold, A. (2000), "Economic and Equity Effects of Marginal Cost Pricing in Transport Case Studies From the European Cities", 71, VATT, Helsinki, Finland.

[20] Giuliano, G. (1994), "Equity and Fairness Considerations of Congestion Pricing", Transportation Research Record, No. 242, pp. 250–279.

[21] Eliasson, J. and Mattsson, L. (2006), "Equity Effects of Congestion Pricing Quantitative Methodology and A Case Study for Stockholm", Transportation Research Part A, Vol. 40, pp. 602–620.

[22] Sumalee, A. (2003), "Optimal Toll Ring Design with Spatial Equity Impact Constraint: An Evolutionary Approach", Journal of Eastern Asia Society for Transportation Studies.

[23] Franklin, P. J. (2006), "A Non-Parametric Analysis of Welfare Redistribution: The Case of Stockholm's Congestion Pricing Trial", Submitted to Transportation Research, Part A: Policy and Practice. December 1, 2005.

[24] Arnott, R., de Palma, A. and Lindsey, R. (1994), "The welfare Effects of Congestion Tolls with Heterogeneous Commuters", Journal of Transport Economics and Policy, Vol. 28, pp. 139–161.

[25] Bhatt, K., Higgins, T. and Berg, T. J. (2008), "Lessons Learned From International Experience in Congestion Pricing" Final report, prepared for U.S. Department of Transportation, Federal Highway Administration.

[26] Giuliano, G. (1992), "An Assessment of the Political Acceptability of Congestion Pricing." Transportation, Vol. 19, pp. 335–358.

[27] Evans, A.W. (1992), "Road Congestion Pricing: When is it a Good Policy?", Journal of Transport Economics and Policy, Vol. 26, pp. 213–243.

[28] Karlstrom, A. and Franklin, J. P. (2008), "Behavioral Adjustments and Equity Effects of Congestion Pricing: Analysis of Morning Commutes During the Stockholm Trial", Transport Research Part A. doi: 10.1016/j.tra.2008.09.008

[29] Ramjerdi, F. (2006), "Equity Measures and Their Performance in Transportation", Transportation Research Record: Journal of the Transportation Research Board, No. 1983, Transportation Research Board of the National Academies, Washington, D.C., pp. 67–74.

[30] Eliasson, J. (2009), "A Cost–Benefit Analysis of the Stockholm Congestion Charging System", Transport Research Part A. doi:10.1016/j.tra.2008.11.014

[31] Small, K. (1983), "The Incidence of Congestion Tolls on Urban Highways", Journal of Urban Economics, Vol. 13, pp. 90–111.

[32] Hau, T. D. (1992), Economic Fundamentals of Road Pricing. Policy Research Working Paper, Transport Division, Infrastructure and Urban Development Department, The World Bank, Washington, D.C.

[33] Gomez-Ibanez, J. A. (1992), The Political Economy of Highway Tolls and Congestion Pricing", In: Federal Highway Administration (Editor), Exploring the Role of Pricing as a Congestion Management Tool. FHWA, Washington, USA.

[34] Flowerdew, A. D. (1993), Urban Traffic Congestion in Europe: Road Pricing and Public Transport Demand, Research Report, The Economics Intelligence Unit, London.

[35] Moses, L.N. and Williamson Jr H. F. (1963), "Value of Time, Choice of Mode and the Subsidy Issue in Urban Transportation." The Journal of the Political Economy, 71, 247–264.

[36] MVA (1995), The London Congestion Charging Research Programme, Final Report, Vol. 1: Text, Government Office for London, HMSO, London.

[37] Langmyhr, T. (1997), "Managing equity: The Case of Road Pricing", Transport Policy, Vol. 4, No. 1, pp. 25–39.

[38] Leeds City Council (2000), Road Pricing in Leeds, Leeds City Council, Leeds.

[39] Transport for London (2002). The Greater London (Central Zone) Congestion Charging Order 2001: Report to the Mayor. Transport for London, London.

[40] Santos, G. and Rojey, L. (2004), "Distributional Impacts of Road Pricing: The Truth Behind the Myth", Transportation, Vol. 31, pp. 21–42.

[41] Horner, W. M. (2004), "Spatial Dimensions of Urban Commuting: A Review of Major Issues and Their Implications for Future Geographic Research", The Professional Geographer, Vol. 56, No. 2, pp. 160–173

[42] Rodrigue, P. J., Comtois C. and Slack B. (2006), The Geography of Transport System, Routledge, New York, NY

[43] O'Kelly, M. and Mikelbank, B. (2002), Social Change and Transportation in US Edge Cities. In Black, W. and Nijkamp, P., (eds), Social Change and Sustainable Transport, Indiana University Press, Bloomington.

[44] Deka, D. (2004), "Social and Environmental Justice Issues in Urban Transportation", in Hanson, S. and Giuliano, G., (eds) The Geography of Urban Transportation, 3rd edition, The Guilford Press, New York, NY, pp. 332–355.

[45] Miller, H. and Shaw, S. (2001), Geographic Information Systems for Transportation: Principles and Applications. Oxford University Press, New York

[46] Dunn, K. (2005), "Interviewing". In Hay, I., (ed.) Qualitative Research Methods in Human Geography). University Press, Oxford, pp. 79–105.

http://dx.doi.org/10.5772/107156

[47] Osterle, A. (2002), Evaluating Equity in Social Policy: A Framework for Comparative Analysis. Evaluation, Vol. 8, pp. 46–59.

[48] Truelove, M. (1993), "Measurement of Spatial Equity", Environmental and Planning C: Government and Policy, Vol. 11, pp. 19–34

[49] Murray, T. A. and Davis, R. (2001), "Equity in Regional Service Provision", Journal of Regional Science, Vol. 41, No. 4, pp. 577–600.

[50] Rescher, N. (1966), Distributive Justice. Bobbs-Merrill, Indianapolis, Indiana

[51] Campbell, A. K. (1976), "Approaches to Defining, Measuring and Achieving Equity in the Public Sector". Public Administration Review, Vol. 36, pp. 556–562.

[52] Daniels, N. (1985), Just Health Care. Cambridge University Press, Cambridge

[53] Rawls, J. (1971), A Theory of Justice. Clarendon Press, Oxford, UK.

[54] Litman, T. (2007), "Developing Indicators for Comprehensive and Sustainable Transport Planning", Transportation Research Record: Journal of the Transportation Research Board, No. 2217, Transportation Research Board of the National Academies, Washington, D.C., pp. 10–15.

[55] PATS, (2000), "Socio-economic principles for price acceptability", Deliverable D2, Status (D). Project funded by the European Commission under the Transport RTD programme of the 4th framework programme.